STUDY GUIDE

VERNON DOMINGO
Bridgewater State College
Bridgewater, MA

---

to accompany

PHYSICAL GEOGRAPHY OF THE GLOBAL ENVIRONMENT

SECOND EDITION

H. J. DE BLIJ
University of South Florida
St. Petersburg

PETER O. MULLER
University of Miami

John Wiley & Sons, Inc.
New York • Chicheter • Brisbane • Toronto • Singapore

Copyright © 1996 by John Wiley & Sons, Inc.

All rights reserved.

Reproduction or translation of any part of this work beyond that permitted by Sections 107 and 108 of the 1976 United States Copyright Act without the permission of the copyright owner is unlawful. Requests for permission or further information should be addressed to the Permissions Department, John Wiley & Sons, Inc.

ISBN 0-471-11636-X

Printed in the United States of America

10 9 8 7 6 5 4 3 2

CONTENTS

****

INTRODUCTION

This study guide is designed to assist you in understanding the material presented in de Blij and Muller's Physical Geography of the Global Environment. The guide is presented in a manner that you can use on your own in tutorial sessions at your institution. The purpose of using the guide is to make your study of physical geography more efficient and meaningful. After working through the textbook and the guide, you should have a deeper knowledge of the many processes in our physical world and also a clearer understanding of the geographer's approach to the study of our planet.

Three primary geography themes are presented as you work through the text and this guide:

1. Locational aspects of physical phenomena (e.g. Where are are hurricanes found? Why are they found there?).
2. Connections between physical elements in the environment (e.g. What impact does climate have on vegetation?).
3. Linkages between human beings and the natural environment (e.g. What are the effects on streams when dams are built?).

As you read the material, continually remind yourself to respond to these themes. They can provide you with an identifiable geographic frame of reference. These themes constitute the primary elements in developing a geographic mindset that allows you to more easily understand the way physical geography is presented. While the material may come from fields such as biology, ecology, geology, or climatology, it is the uniquely ***geographic perspective*** which sets it apart. Geography incorporates ideas from many different disciplines, all the while approaching knowledge in an integrative, spatial manner that uses distinctive skills. It is important to remember that a major goal of the course you are taking is the transmission of information about earth environments through the development of your geographic vision. Your ability to "see" geographically and to make connections among phenomena and between people and the natural world depends on how well you learn and practice the skills of graphicacy and hypothesis developing that are emphasized in this study guide.

GRAPHICACY

Your task in studying geography is made much easier when you begin to think similar to geography authors and instructors. Thinking geographically is essential if you are to grasp the explicit and the implicit content of the material. The geographer "sees" the landscape in a particular way which we shall refer to as **geo-graphicacy**, the communication of spatial information by means of visual data sources such as maps, diagrams, photographs, and charts. The wealth of information found in cartography and photography (both using the root "graph") needs to be recognized and interpreted as a central part of geographical analysis and communication. The other modes of communication are literacy (reading/writing skills), numeracy (number appreciation/ manipulation), and articulacy (listening/speaking).

Graphicacy, is a necessary skill in interpreting the many facts and figures presented in your geography text. The skill of graphicacy involves the ability to "read the landscape" and to discern relationships between otherwise separate features in the environment. In order to see the world as a geographer, one needs to appreciate the many spatial relationships which exist. Spatial in this context means the locational or distributional aspects of phenomena - the **WHERE** and **WHY** of things.

Taking one example from physical geography, we note in the following sketch that while the left side of the mountain is moist, the right side receives very little rain.

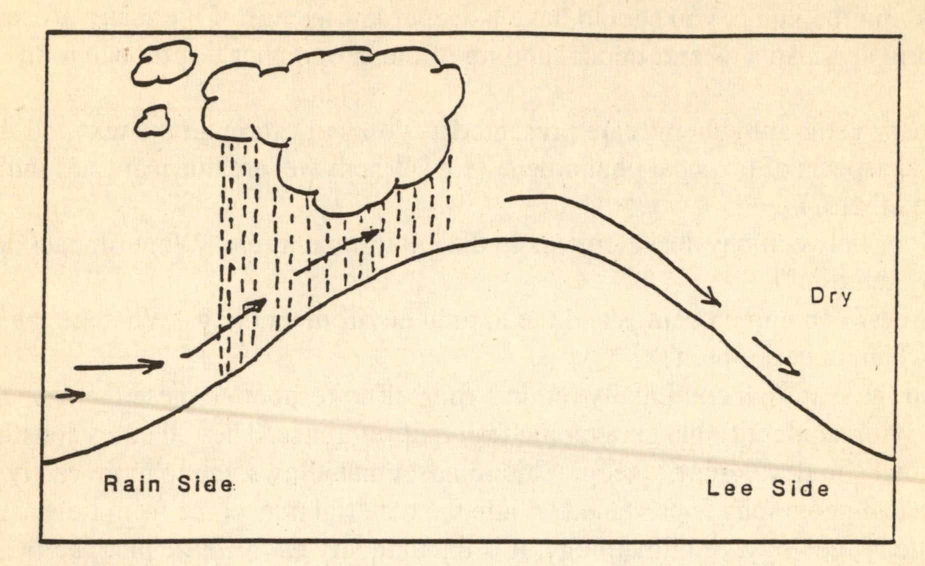

Figure 1. Rain and lee-side

When we view the interaction of the various elements - wind, mountain, location - we can deduce that the mountains serve as a barrier to rain-bearing wind. Most of the rain is deposited on the upwind side of the mountain, so that by the time these winds reach the leeside, the winds are dry and they create very different climates. It is the relative location of A and B (that is, relative to the mountain) that determines whether they receive more or less rainfall. Figure 1 enables us to grasp rapidly the nature of the relationship of dry area to mountain. While words may convey this very same idea, the figure best communicates the relationship under discussion. Just as the architect's plan of a house gives us a clear sense of layout and relationship in distance and accessibility between kitchen and bedroom, so maps and diagrams show not only what exists, but how each element is connected to all the others. We are able to go beyond just isolated facts and develop questions about the effect of one feature on the other - which carpet will be the most worn, that between kitchen and bedroom or between living room and kitchen? The maps and diagrams in the text and in this study guide are an integral part of the course content. They must be examined, interpreted, and applied if you are to develop your geographic "eye".
While most of the focus for geographers is on spatial aspects of the landscape, there is also a concern with the element of ***time***. It is a given that nothing stays quite the way it is and it therefore becomes important to pay attention to the process that changes features of the environment from one time to the next:

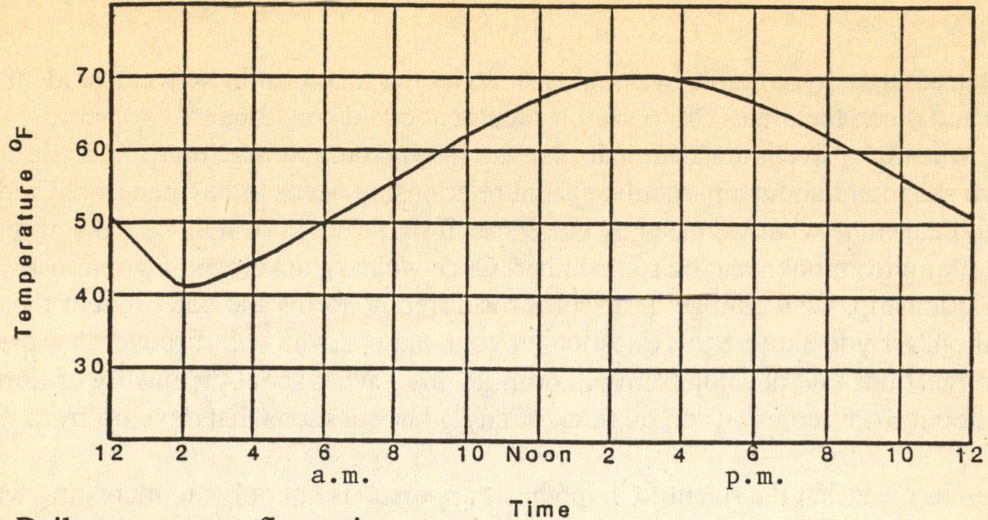

Figure 2. Daily temperature fluctuation

In this example, the changes in temperature over the course of a 24-hour period can best be illustrated graphically. This graph display conveys the nature of the relationship between temperature and time change in a manner that allows us to explore the nature of that relationship. Why do we find the highest temperature at 2:00 p.m. and not at noon as could be expected?; and what about a 2:00 a.m. lowest temperature? These are some of the types of questions that graphicacy presents. They sharpen our ability to interpret the environment and to develop and explore fresh questions about the nature of the world around us.

Graphicacy is a skill that enables you to develop greater insight into the subject, making your studying much more productive. Throughout this course, be sure to pay particular attention to the many maps, sketches, diagrams, and charts that are presented throughout the book. They have been included in order to make learning much easier and much more geographic.

HYPOTHESIS CONSTRUCTION

In this course you will learn a great many new (and longstanding) facts about the world around us. It is important that you know these definitions, descriptions, and numerical values because they form the key elements for knowing about and caring for the planet. But as a scientific endeavor, the concerns are not only with the nature of a particular phenomenon, but especially with the ***relationship*** between those phenomena. Rather than being merely descriptive, the attempt here will be to develop and examine the nature of those relationships. Knowledge of how one element affects the other helps us develop tentative predictions of what to expect when these elements are present. Such a statement about the direction of the relationship between phenomena is called a ***hypothesis***. While the statement is tentative, it does serve as a guide as we continue exploring the complexities of the world. Hypotheses are not a new feature; they occur in everyday life. When you ask why many students do better in some classes than in others you are constructing a set of hypotheses to explain a particular phenomenon. Once stated, the hypothesis provides a framework within which you can explore causes and consequences in the world around us.

To begin constructing a hypothesis, we could ask about the relationship between student grade performance and dorm location. Our research statement could be: "Students do better academically when they live in resident halls that are located close to thelibrary".

This tentative statement about a particular spatial relationship needs to be operationalized. We would have to determine what we mean by closeness, if that were in straight-line distance or in time spent walking from one location to the other. Once we have developed a good map showing our spatial relationship, we would get test scores for different dorms and determine if there is some correlation. As you explore this question on your campus, you will discover that there are many further questions that develop from our original one - What about the quality of library time? What about SAT scores? Can you think of any other questions that develop from this research?

It soon becomes clear that the scientific hypothesis approach is not only an interesting way of studying any issue but it also provides a strong, clear focus to our explorations. The scientific approach presented thus far is made of four parts: observation, questions about what is observed, identification, and statement of the problem in the form of a testable hypothesis. The evaluation of the hypothesis means that it can either be confirmed or rejected. Whatever the outcome, developing and stating the hypothesis has helped us make sense of what could otherwise be a confusing mix of data.

In many of the units of this study guide you will be called upon to develop such hypotheses and to evaluate them. The text is a valuable tool in this process, so be willing to flip ahead or back to uncover the relationships we are exploring. Everything is related - now let us find our just **HOW** they are connected.

And lastly. Welcome to geography. The subject can provide some of the answers for the environmentally conscious person who is concerned with understanding and caring for the planet. Good luck as you explore these global environments.

****

On using the guide: Each text unit is presented through a similar sequence of learning activities. The guide invites you to become involved in the course material by selecting answers, writing responses to questions, drawing sketches, and interpreting graphical information. I recommend that you work systematically through each unit as many of the early concepts are used as a foundation for later discussion. The answer key for some questions is on page 349.

Suggestions:

Write out your answers as this activity will help consolidate the information. Text page numbers are supplied with many of the questions. Only use these to check your answers.

Use a pencil with most of the guide questions, especially with the graphicacy section.

Draw those sketches. The sketches you are asked to construct are an essential part of learning geography. Remember you will not be evaluated as if you were a van Gogh. It is more important that you grasp the nature of the geographic relationships under discussion. Good luck and enjoy the course.

PART ONE

A GLOBAL PERSPECTIVE

Unit 1

Introducing Physical Geography

(A) SUMMARY AND OBJECTIVES

If physical geography is about anything, it is about DISCOVERY. The excitement of the field of physical geography becomes evident when we note the many voyages of exploration taken by geographers through the ages. This unit reveals the efforts of explorers such as Alexander von Humboldt (in South America) and John Wesley Powell (in the Grand Canyon) who were interested in discovering how the earth was and still is being created. This search for explanation is at the root of our voyage in this course. We will build on the ideas developed by these early explorers to examine the processes at work today that change conditions in the atmosphere and in the earth's rock structure. A significant fact to remember here and throughout the text, is that while geography has much in common with other subject areas, its distinctive character lies in its integrative nature, especially in closely relating human beings to the environment.

After reading this unit you should be able to:

1. Understand how physical geography fits into the overall field of geography.
2. Describe the relationship between geography and other disciplines.
3. Identify the elements of the systems approach.
4. Understand the nature of theoretical models.

(B) KEY TERMS AND CONCEPTS

climatology
dynamic equilibrium
feedback
geography
geomorphology
meteorology
model
orders of magnitude
physiography
spatial systems

(C) MULTIPLE CHOICE

_____ 1. Pedology is the study of : (p. 8)
(a) weather
(b) soils
(c) plants

_____ 2. What do we call a feedback mechanism that operates to keep a system in its original condition? (p. 10)

(a) positive feedback
(b) a passing grade in geography
(c) negative feedback

______ 3. What was the major contribution that William Morris Davis made to the field of geography? (p. 9)

(a) discovering new regions of the world
(b) describing ocean currents
(c) analyzing how rivers wear down rock structures

______ 4. What is geodesy the study of? (p. 6)

(a) the depth of the oceans
(b) size and shape of the earth
(c) domes and maps

______ 5. What do we call the boundaries between systems? (p. 10)

(a) fronts
(b) interfaces
(c) contacts

(D) TRUE OR FALSE

1. Geographers are interested in chronology while historians are interested in problems involving space. (p. 5)
T _____ F _____

2. Geomorphology focuses on the processes of land formation while physiography is concerned with the regional results. (p. 8)
T _____ F _____

3. A city is an open system. (p. 10)
T_____ F_____

4. The word "geography" was first coined in the 19th century. (p. 5)
T _____ F _____

5. Eratosthenes defined geography as a description of the earth: (p. 5)
T_____ F _____

(E) SHORT-ANSWER QUESTIONS

1. What are the two approaches that sets geography apart from the other scholarly disciplines? (p. 5)

____

____

2. What are the three main questions which geography poses? (p. 5)

____

____

3. What are the two main tasks of a **cartographer**? (p. 6)

____

4. Explain why von Humboldt's contributions to geography should be described as natural geography rather than physical geography. (p. 7)

____

____

5. How does soil pedology differ from soil geography? (p. 8)

____

____

6. What is meant by a **system**? (p. 10)

____

7. Distinguish between an open and a closed system. (p. 10)

____

8. Describe the nature of positive feedbacks. (p. 10)

____

9. How do **models** help us study aspects of our world? (p. 11)

____

____

10. Why do geographers use different orders of magnitude in their work? (p. 11)

____

____

(F) MATCHING

Match the following terms and their meanings:

_______ 1. geomorphology
_______ 2. geodesy
_______ 3. system
_______ 4. physiography
_______ 5. climatology
_______ 6. dynamic equilibrium
_______ 7. biogeography

a. study of climates and their distribution
b. a system in operation, but neither growing nor getting smaller
c. study of landforms and their processes
d. the study of distribution of plants and animals
e. study of the size and shape of the earth
f. a set of related events or objects and their interactions
g. study of the regional results of landform formation

(G) ESSAY QUESTIONS

1. Describe what is meant by geography's role as a *synthesizing subject.*

2. What natural feature in your region would be of interest to a physical geographer?

3. Describe two research questions that a physical geographer might ask about this feature..

4. Describe how the focus of physical geography has changed in the last 200 years.

5. What future changes do you think await the ever changing field of physical geography?

6. How did the work of Alexander von Humboldt serve to formalize the field of physical geography?

7. What is meant by a ***model*** in scientific research?

8. Describe the nature of the ***systems approach*** in physical geography.

(H) GRAPHICACY

1. Describe the nature of human/land interaction that can be inferred from the photograph of Miami Beach in Figure 1-6. on page 11 of the text.

(I) HYPOTHESIS CONSTRUCTION

1. Examine the photograph of the Grand Canyon in Figure 1-5 on page 9 and describe this landform.

2. Name three different elements that are shown in the photograph that could have been responsible for creating the Grand Canyon features.

3. Complete the following:
The Grand Canyon could have been created by ______________________________

& ______________________________

You now have 4 ***variables*** - something that is thought to influence or be influenced by something else. They are called variables because they can change, depending on the impact of other phenomena. But not all variables are used in the same way. Grand Canyon is the dependent variable, because its existence is dependent on the others. The other 3 are independent variables because they are influencing changes in the Grand Canyon. One could write this out as:

The dependent variable (DV) is determined by an independent variable (IV) plus another independent variable (IV) plus a third independent variable (IV). Or we could say:

$$DV = IV + IV + IV$$

A hypothesis is a statement about the relationship between one variable and the others. Terms like "influenced by", "determined by", "associated with", "increases with" are often used in hypotheses to show relationship between variables.

Now write a hypothesis that focuses on what you can see in the Grand Canyon photograph.

Congratulations! You're now on your way to scientific research.

Unit 2

The Planet Earth

(A) SUMMARY AND OBJECTIVES

Ever wonder why, if the earth is largely covered with water, we do not refer to the planet as Aqua? Consideration of this question leads us to the core issue in geography - namely that our study of the physical earth cannot easily avoid including humans and their actions. Both cultural and natural elements need to be considered together if we are to make sense of the world around us. Our own mental maps of the planet will be challenged here as we consider the global distribution of people, land, and water. What is the largest continent and how does this influence weather and climate? This unit addresses our life spheres and examines how many there are of us on this orb, where we are distributed, and what our future numerical growth could entail.

After reading this unit you should be able to:

1. Describe the different spheres of the earth system.
2. Understand the interaction between these spheres.
3. Identify the features of the seafloor topography.
4. Explain the nature of population doubling time.

(B) KEY TERMS AND CONCEPTS

abyssal plain
atmosphere
biosphere
continental slope
continental shelf
ecumene
hemisphere
hydrosphere
lithosphere
midoceanic ridge
population doubling time
seamount

(C) MULTIPLE CHOICE

1. What percentage of the surface of the earth is covered by water or ice? (p. 17)
   (a) 54%
   (b) 63%
   (c) 71%

2. What is the approximate 1995 world population? (p. 18)
(a) 4.2 billion
(b) 5.7 billion
(c) 7.3 billion

3. Which continent is known as the "plateau continent"? (p. 17)
(a) Africa
(b) South America
(c) Eurasia

4. What percentage of the world's population lives in the northern hemisphere? (p. 18)
(a) 60%
(b) 75%
(c) 90%

5. How many human beings are added to the world's population every day? (p. 18)
(a) 130,000
(b) 420,000
(c) 265,000

(D) TRUE OR FALSE

1. The southern hemisphere contains about 30% of the earth's land area. (p. 14)
T _____ F _____

2. Eurasia is the largest landmass on earth. (p. 17)
T _____ F _____

3. North America is larger than South America. (p. 17)
T _____ F _____

4. The continental shelf is the largest expanse of lower relief ocean floor. (p. 21)
T _____ F _____

5. Most of the world's population live far away from the seacoasts. (p. 18)
T _____ F _____

(E) SHORT ANSWER QUESTIONS

1. Of what does the **biosphere** consist? (p. 13)

2. Describe the nature of a regional subsystem. (p. 13)

3. Name four differences between the northern and southern hemispheres. (p. 14)

4. Name the six continental landmasses. (p. 17)

5. How does the Mediterranean Sea differ from the Caribbean Sea? (p. 16)

6. Describe the differences between the north and south polar areas. (p. 14)

7. What impact does North America's topography have on its weather and climate? (p. 17)

8. What is the **continental margin** and what are its three components? (p. 21)

9. How does the continental slope differ from the continental rise? (p. 21)

10. What are **mid-oceanic ridges** and why are they important in our understanding of the earth's physical geography? (p. 21)

____

____

(F) MATCHING

Match the following terms and their meanings:

_______	1. ecumene	a. the water segment on the earth and in the atmosphere
_______	2. abyssal plain	b. sloping, submerged plain at the edge of continents
_______	3. hydrosphere	c. the livable world where human settlement is possible
_______	4. seamount	d. large expanse of lower relief ocean floor
_______	5. atmosphere	e. zone of life
_______	6. continental shelf	f. volcanic mountains on the ocean floor
_______	7. biosphere	g. blanket of air adhering to the earth's surface

(G) ESSAY QUESTIONS

1. Describe the four spheres as presented in the opening photograph on page 12 of the text.

2. Explain how the four spheres in this photograph interact.

3. In a short paragraph, relate the distribution of population in Asia on Figure 2-5 to the map of continental topography (Figure 2-6).

(H) GRAPHICACY

1. One revealing measure of population growth is doubling time, or the time it takes for the population to double. Use the table of doubling time to construct two graphs, one showing increases in population, and the other illustrating doubling time.

Year	Population	Doubling time (years)
1	250 mil	
1650	500 mil	1,650
1850	1,000 mil	200
1930	2,000 mil	80
1975	4,000 mil	45
1992	5,420 mil	41

Population (millions)

7000
6000
5000
4000
3000
2000
1000

500 1000 1500 2000

Year

(i)

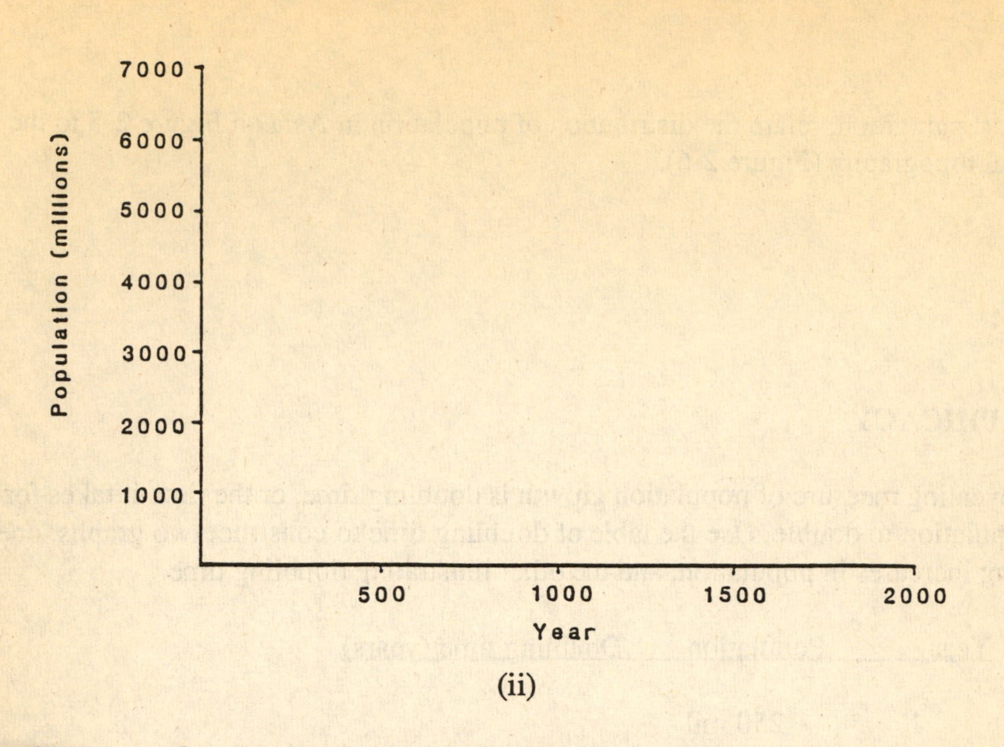

(ii)

Describe the nature of population change in (i) and (ii).

(I) HYPOTHESIS CONSTRUCTION

1. Based on the discussion of hypotheses in the introduction and in unit 1, write out a hypothesis statement that follows our format.

$$DV = IV + IV + IV$$

(Population growth should be your dependent variable).

2. Which of your independent variables is the most important? Explain.

3. Based on the data provided in the unit, what would you predict the world population would be in 2030. Explain your reasoning.

Unit 3

Mapping The Earth's Surface

(A) SUMMARY AND OBJECTIVES

Can you imagine a world without maps? A world where we could not be sure where things were located? The maps that help us get around campus or to a neighboring town are so much an integral part of our lives that we hardly pause to consider that these maps were human inventions that enable us to deal with our environment. While simple maps have their origin in prehistoric times, today we deal increasingly with sophisticated maps in the newspaper, in magazines, or on a computer screen. This unit asks us to pause and consider the various elements of maps - distance, direction, scale, projection, and symbolization - that together remind us that to study geography is to study maps. The saying "without geography you're nowhere" is largely true because of the regularity of map use by geographers. Maps are there for you to use and to enjoy.

After studying this unit you should be able to:

1. Understand and explain the nature of maps.
2. Explain the reference system used in maps.
3. Distinguish between the different types of projections.
4. Discuss the latest mapmaking technology.

(B) KEY TERMS AND CONCEPTS

cartography	Mercator projection
contours	meridians
geographic information system	parallels
isohyets	remote sensing
latitude	scale
longitude	standard parallel

(C) MULTIPLE CHOICE

_____ 1. What area was depicted on the oldest surviving maps? (p. 24)

(a) California
(b) China
(c) Mesopotamia

_____ 2. How many degrees does a circle have within it? (p. 24)

(a) 90
(b) 360
(c) 180

_____ 3. What is the length in kilometers, of one degree of latitude at the equator? (p. 25)

(a) 200 km
(b) 102 km
(c) 111 km

_____ 4. On which end of the electromagnetic spectrum do we find radio waves? (p. 35)

(a) short wave
(b) long wave
(c) medium wave

_____ 5. What variable is connected by isohyets? (p. 33)

(a) slamdunks
(b) precipitation
(c) altitude

(D) TRUE OR FALSE

1. Lines of longitude are also known as parallels. (p. 24)
T _____ F _____

2. Planar projections transfer the geographic grid to a cone. (p. 27)
T _____ F _____

3. Isohyets are lines joining places of equal precipitation. (p. 32)
T _____ F_____

4. The best known example of a planar projection is the Mercator projection.. (p. 26)
T _____ F _____

5. When a map projection is said to be **conformal**, it is preserving the shape of an area. (p. 26)
T _____ F _____

(E) SHORT-ANSWER QUESTIONS

1. According to geographers, what is the most common form of conveying information about location and place? Why is this such a significant mechanism for understanding? (p. 23)

____

____

2. Define **cartography**. (p. 24)

____

3. Name the three divisions in each degree? (p. 24)

____

4. How was the first idea of direction developed? (p. 24)

____

5. What is meant by the **geographic grid**? (p. 24-25)

____

6. Name three ways in which parallels and meridians differ from each other. (p. 25)

____

____

7. What is the **standard parallel** and what is its role? (p. 26)

____

8. Name four different types of map symbols and give an example of each. (p. 30-32)

____

9. How does **GIS** differ from previous map-making approaches? (p. 34)

____

10. Describe the difference between passive and active systems used in remote sensing. (p. 36)

____

____

(F) MATCHING

Match the following terms and their meanings:

_______ 1. scale		a. the 0^0 north-south line
_______ 2. isohyet		b. arrangement of earth's geographic grid on a flat surface
_______ 3. parallels		c. lines connecting places having the same value of a phenomenon
_______ 4. prime meridian		d. north-south lines on a map
_______ 5. contours		e. a line of true and compass bearing
_______ 6. projections		f. ratio of the size of an object on a map to its real world size
_______ 7. isolines		g. east-west lines on a map
_______ 8. rhumb lines		h. lines showing equal height above sea-level
_______ 9. longitude		i. isolines connecting places of equal precipitation

(G) ESSAY QUESTIONS

1. Refer to a globe to answer this question. Could the position of the equator be changed? Explain.

2. Again, refer to a globe to answer this question. Could the position of the 0^0 longitude line be changed? Explain.

3. Find three maps that cover your region; one should be a world map, another should be a map of the United States, and a third a map of your town. Be sure each map has a scale.

a. Write down these respective scales in ratio form.

b. Compare the use of each of the differently scaled maps.

4. Give the geographic coordinates (latitude and longitude) to the degree, minute, and second for your college (you will need to consult a topographic map of your area).

5. Why is the latitude value always given first when geographic coordinates are provided?

6. Compare the features of the world maps (Figures 3-5 and 3-7). Refer to your textbook and determine for what purposes each would be most appropriate; find at least two examples of each in the text.

7. Ratio scale is given as 1: --- ; if the suggested scale is 1 inch represents 400 feet, what is the correct form for the ratio scale (show all calculations).

8) Compare the characteristics of map projections, by completing the following table:

Projection	distortion	grid lines	tangency
Cylindrical			
conic			
planar			
equal-area			

9) Draw two cross-sections of the contour map, X to Y in Figure 3-12; use a vertical scale of 1"=100' in the first and 1"=200' in the second.

(i)

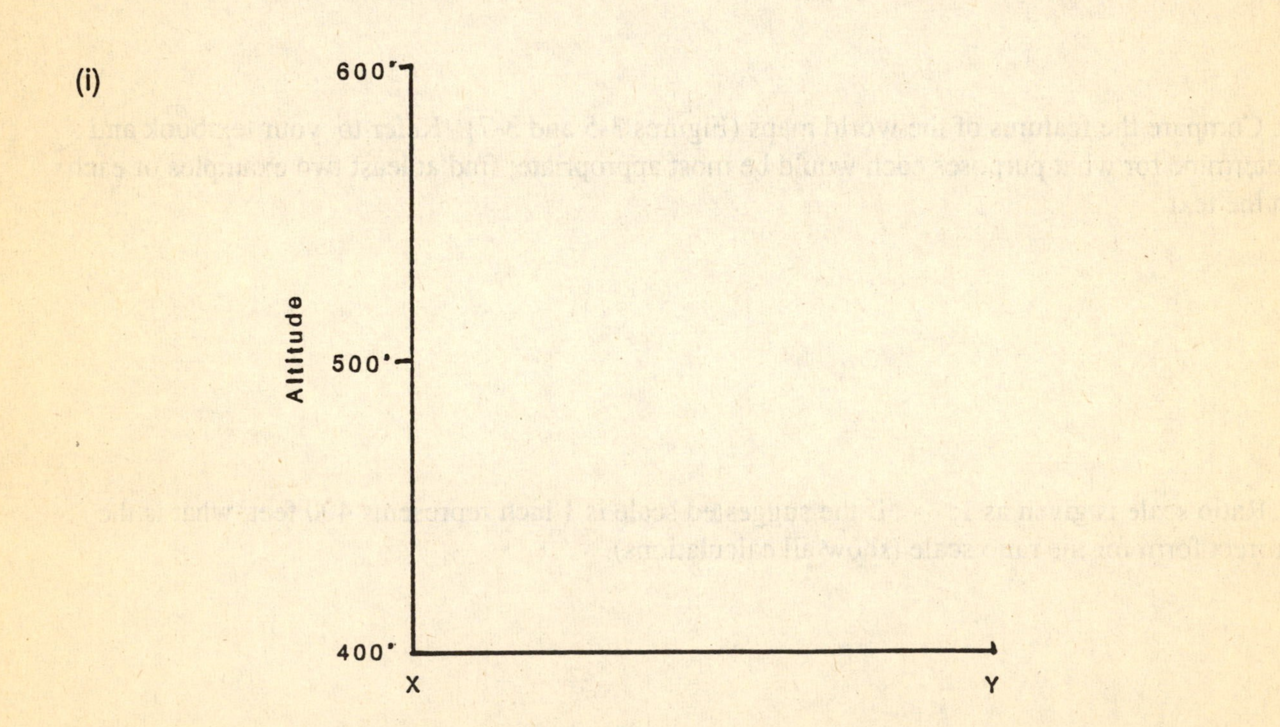

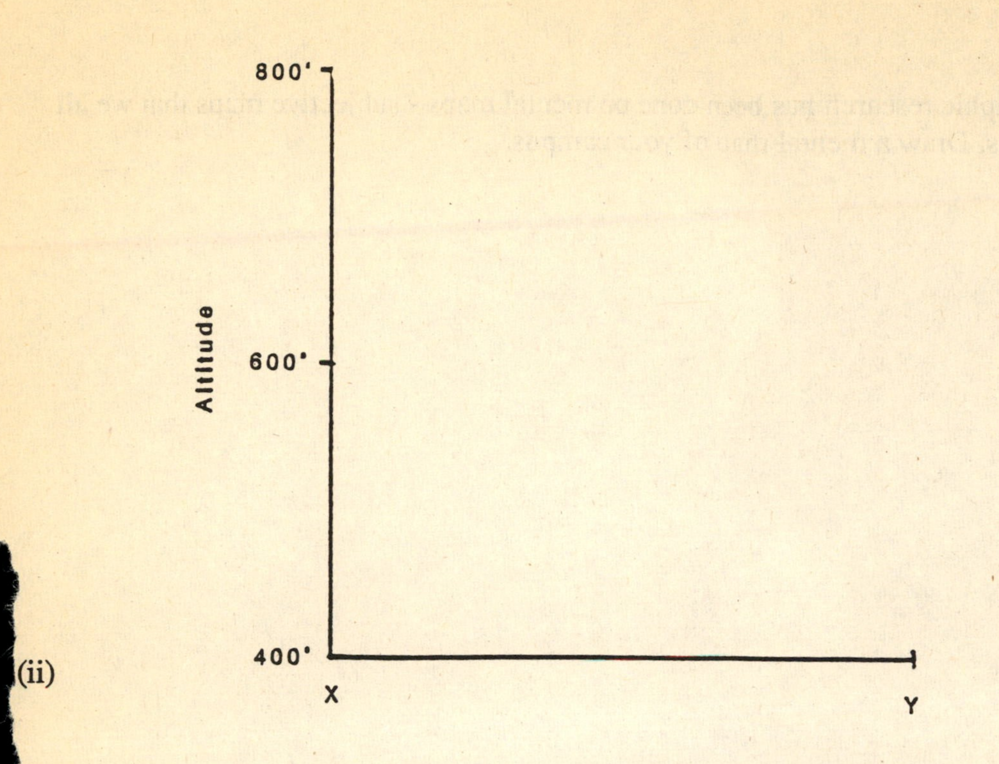

(ii)

). Discuss the differences produced by using these different vertical scales.

I) GRAPHICACY

Discuss the extent to which the map in the opening photo on page 23 of the text differs from the map n Figure 2-5. Explain these differences.

2. A great deal of geographic research has been done on mental maps - subjective maps that we carry around in our heads. Draw a mental map of your campus.

(I) HYPOTHESIS CONSTRUCTION

1. What factors may be responsible for the pattern of the distribution of high temperatures in Figure 3-10 of the text?

2. Write a hypothesis statement to explain the pattern in Figure 3-10.

3. Find three maps that cover your region; one should be a world map, another should be a map of the United States, and a third a map of your town. Be sure each map has a scale.

a. Write down these respective scales in ratio form.

b. Compare the use of each of the differently scaled maps.

4. Give the geographic coordinates (latitude and longitude) to the degree, minute, and second for your college (you will need to consult a topographic map of your area).

5. Why is the latitude value always given first when geographic coordinates are provided?

6. Compare the features of the world maps (Figures 3-5 and 3-7). Refer to your textbook and determine for what purposes each would be most appropriate; find at least two examples of each in the text.

7. Ratio scale is given as 1: --- ; if the suggested scale is 1 inch represents 400 feet, what is the correct form for the ratio scale (show all calculations).

8) Compare the characteristics of map projections, by completing the following table:

Projection	distortion	grid lines	tangency
Cylindrical			
conic			
planar			
equal-area			

9) Draw two cross-sections of the contour map, X to Y in Figure 3-12; use a vertical scale of 1"=100' in the first and 1"=200' in the second.

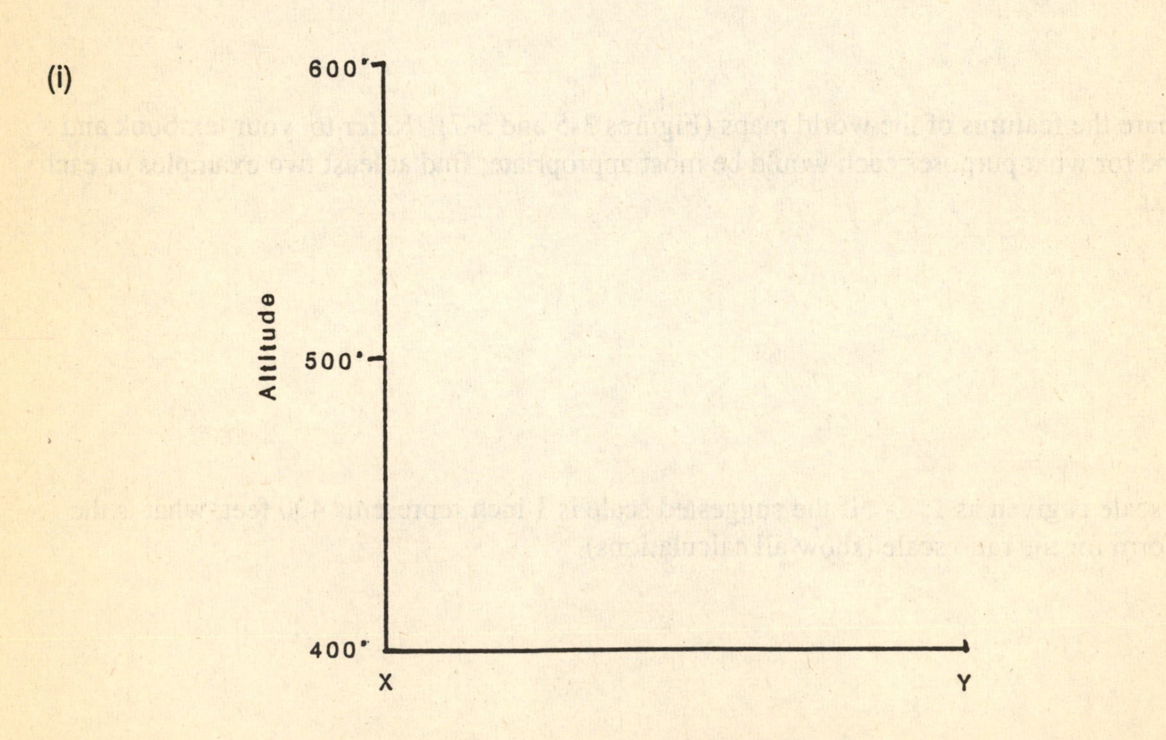

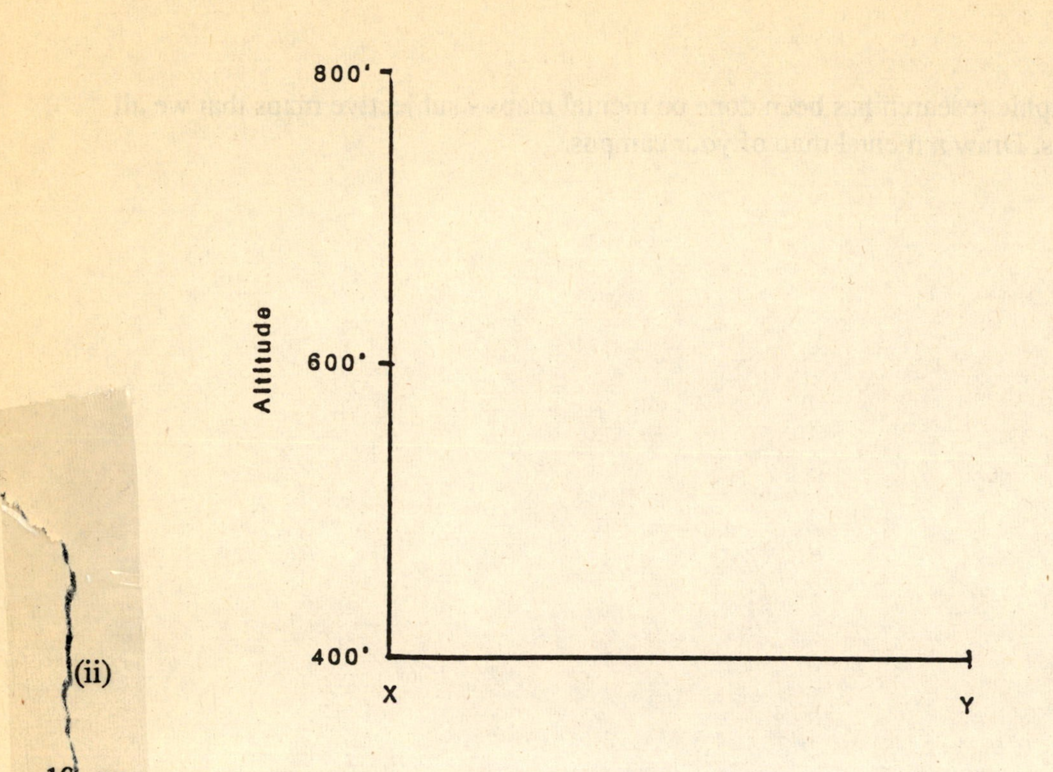

(ii)

10. Discuss the differences produced by using these different vertical scales.

(H) GRAPHICACY

1. Discuss the extent to which the map in the opening photo on page 23 of the text differs from the map in Figure 2-5. Explain these differences.

2. A great deal of geographic research has been done on mental maps - subjective maps that we all carry around in our heads. Draw a mental map of your campus.

(I) HYPOTHESIS CONSTRUCTION

1. What factors may be responsible for the pattern of the distribution of high temperatures indicated in Figure 3-10 of the text?

2. Write a hypothesis statement to explain the pattern in Figure 3-10.

Unit 4

The Earth in the Universe

(A) SUMMARY AND OBJECTIVES

"A small step for man, a giant step for mankind" - with these words in 1969 our horizons as a species were immediately widened to a world beyond our own planet. When the first human walked on the moon, it was an extension in geographic knowledge and also a triumph in a voyage of exploration that our earlier ancestors could not have imagined. What was once science fiction may become everyday occurrences even during our lifetime. It is therefore important that we get to know about our nearest neighbors - the other eight planets, their moons, and our sun. Tomorrow may be the day when we go to worlds where "no one has gone before".

After reading this unit you should be able to:

1. Understand the basic units of the universe.
2. Describe the position of planet earth in the larger context of our solar system.
3. Explain the role of solar activity and its effect on our planet.
4. Contrast the characteristics of different planets.

(B) KEY TERMS AND CONCEPTS

galaxy	planetesimal
gravity	revolution
light-year	rotation
Milky Way	solar system
planet	universe

(C) MULTIPLE CHOICE

______ 1. How long does it take light to travel from the sun to the earth? (p. 40)
(a)15 minutes
(b) 8 minutes
(c)10 minutes

______ 2. Approximately when did the Big Bang occur? (p. 41)
(a) 10 billion years ago
(b) 18 billion years ago
(c) 15 billion years ago

______ 3. What are the main gases in the sun? (p. 42)
(a) hydrogen and helium
(b) nitrogen and oxygen
(c) helium and nitrogen

______ 4. How long is the cycle of magnetic storm activity on the sun? (p. 42)
(a) 5 years
(b) 7 years
(c) 11 years

______ 5. Which of these planets has the hottest surface? (p. 44)
(a) Earth
(b) Mars
(c) Venus

(D) TRUE OR FALSE

1. Our sun is only about 4.6 billion years old. (p. 42)
T _____ F _____

2. The inner planets consist of Mars, Venus, and Satur. (p. 42)
T _____ F _____

3. The planet Pluto was only discovered in 1930. (p. 42)
T _____ F _____

4. Venus is often called the Evening Star. (p. 45)
T _____ F _____

5. The earth receives less than one-billionth of the light and heat given off by the sun. (p. 42)
T _____ F _____

(E) SHORT-ANSWER QUESTIONS

1. Explain what is meant by the Big Bang theory. (p. 40)

____

____

2. Define **gravity**. (p. 41)

____

3. Explain the difference between planets and stars. (p. 42)

____

____

4. Explain the differences between comets and meteoroids. (p. 42)

____

____

5. Which two planets cannot be seen with a telescope? (p. 43)

____

6. Describe the three physiographic categories of the moon's surface. (p.47)

____

____

7. Describe the history of the moon. (p. 47)

____

____

8. Define a **light-year**. (p. 40)

____

____

9. What is the significance of the "Red Spot" on Jupiter? (p. 46)

____

____

10. Describe the **Milky Way**. (p. 40)

____

____

(F) MATCHING

Match the following terms and their meanings:

_______ 1. solar system

_______ 2. comets

_______ 3. galaxy

_______ 4. Jovian

_______ 5. asteroids

_______ 6. a revolution

a. an organized assemblage of billions of stars

b. resembling the planet Jupiter

c. a large belt of small planetesimal-like materials

d. one complete circling of the sun

e. the sun, its planets, and related residual materials

f. small bodies of frozen gases and related materials

(G) ESSAY QUESTIONS

1. Explain what would happen if the output from the sun were reduced by one-third.

2. Why are the inner planets referred to as **terrestrial**?

3. Explain how Earth and Venus are similar.

4. Explain how Earth and Venus are different.

(H) GRAPHICACY

1. Examine Figure 4-9 and answer the following question:
Which climatic conditions would need to be present if we (Earthlings) were to colonize this area.

2. After examining Table 4-1 list the three planets that come closest to conditions on earth.
Explain your choices.

(I) HYPOTHESIS CONSTRUCTION

1. Develop a hypothesis to explain why planet Earth is better suited to human occupation than any of the other planets.

Unit 5

Earth-Sun Relationships

(A) SUMMARY AND OBJECTIVES

It is easy to understand why so many early humans worshiped the sun (some people still do). We would of course be nothing without the sun which provides us with our light and energy. It is our primary relationship, the one which determines our climate, our plant activity, and the conditions for the big football game. In many ways, we can say that we are what we are because of the sun. Secondarily, we are what we are because of where we are located on the earth's surface relative to the rays of the sun. The variation of solar radiation from the equator to 45^0N and to the poles is directly a result of the earth's axis and of the resulting differences in length of day. We may not all worship the sun, but we must surely appreciate its role in our existence.

After studying this unit you should be able to:

1. Understand the various movements of the earth.
2. Explain the seasonal impact of the earth-sun relationship.
3. Calculate standard time.
4. Describe possible climate changes with a change in the angle of the earth's axis.

(B) KEY TERMS AND CONCEPTS

angle of incidence	plane of the ecliptic
aphelion	rotation
equinox	solstice
insolation	tropics
perihelion	zenith

(C) MULTIPLE CHOICE

____ 1. What is the plane of the ecliptic? (p. 53)
(a) when the moon moves behind the sun
(b) the line joining the poles
(c) the line of the earth's orbit

____ 2. On what date does the North Pole receive 24 hours of daylight? (p. 53)

(a) March 21
(b) June 22
(c) December 23

____ 3. If the time is 9 am in New York City, what is the time in London? (p. 55)

(a) 1 pm
(b) 2 pm
(c) 4 am

_____ 4. What is the term used when explaining that the earth's axis maintains itself during the entire revolution? (p. 53)

(a) parallelism
(b) congruence
(c) regularity

_____ 5. Between which latitudes is the sun always to the south of the zenith? (p. 57)

(a) 23 1/2^{0}N and 90^{0}N
(b) 47^{0}N and 90^{0}N
(c) 23 1/2^{0}S and 90^{0}S

(D) TRUE OR FALSE

1. The earth rotates from east to west. (p. 51)
T _____ F _____

2. On December 22 the vertical rays of the sun will be at the Tropic of Capricorn. (p. 53)
T _____ F _____

3. Fifteen degrees of longitude equals a one hour change in standard time. (p. 56)
T _____ F_____

4. Aphelion is the position of earth and sun on January 3rd. (p. 51)
T _____ F _____

5. If the earth's axis was not tilted, the planet would constantly be in a position of equinox. (p. 56)
T _____ F _____

(E) SHORT-ANSWER QUESTIONS

1. Why is the sun's heat unevenly distributed in space and time across the earth's surface? (p. 51)

2. Describe the shape of the earth's orbit around the sun. (p. 51)

3. Explain the difference in rotational speeds at the equator and at 40^0N. (p. 51)

4. Describe the nature and effect of **Coriolis force**. (p. 51)

5. Why is it technically wrong to speak of the sun "rising" or "setting"? (p. 52)

6. What is the northern hemisphere vernal equinox and when does it occur? (p. 54)

7. Define the **circle of illumination**. (p. 56)

8. What are the two main factors which determine the annual insolation received at a point on the earth's surface? (p. 57)

9. What social and technological changes lead to the standardization of earth time? (p. 55)

____

____

10. Describe the function of Daylight-Savings Time. (p. 56)

____

____

(F) MATCHING

Match the following terms and their meanings:

_______	1. oblate spheroid	a.	position when the earth is closest to the sun
_______	2. parallelism	b.	force of deflection created by the rotation of the earth
_______	3. insolation	c.	earth-sun positions on March 21 and September 23
_______	4. solstice	d.	the axis of the earth remaining parallel to itself during earth orbit
_______	5. Coriolis force	e.	the earth's shape, bulging and flattening
_______	6. perihelion	f.	the earth-sun position on June 22 and December 22
_______	7. equinox	g.	incoming solar radiation

(G) ESSAY QUESTIONS

1. How does the specific angle of the earth's axis determine our weather and climate?

2. If the earth's axis angle were increased by 10^0F, what climate change would be experienced at:

a. the equator

b. the poles

(H) GRAPHICACY

1. Complete the sketch of the international date line, indicating the deviations from 180^0, the United States and Asia, and the time changes when passing from one hemisphere to the other.

180^0

2. Name the standard time lines from which the following cities take their time:

Boston, Massachusetts ______________

Cairo, Egypt ______________

Tokyo, Japan ______________

3. If it is 5:00 p.m. Saturday in Boston, Massachusetts, what is the time in Cairo, Egypt? First draw a sketch to show the relationships over Greenwich and then show all your calculations.

0°

4. If it is 5:00 p.m. Wednesday in Chicago(90°W), what is the time in Tokyo(135°E)? Cross over the International Date Line; draw a sketch and show all calculations.

180°

(I) HYPOTHESIS CONSTRUCTION

1. Write down your present latitude to the nearest minute (use the geographic grid index of an atlas or a topographic map of your area).

2. After examining Figure 4-4, draw a time-lapse sketch of the daily sun position for your latitude for two dates: June 22 and December 22. (Use a solid line for June and a dashed line for December).

____

3. Write down an statement explaining the variation between the seasons as seen in your sketch in 2 above.

PART TWO

ATMOSPHERE AND HYDROSPHERE

Unit 6

Composition and Structure of the Atmosphere

(A) SUMMARY AND OBJECTIVES

The "sea of air" in which we find ourselves every day is just right for us and for our survival, but are we now changing it for the worse? While change has always been a constant factor on the earth, our concern today has shifted to the dramatically increased rate of change which industrialization has brought about. This is the only planet we have and as such we need to develop an understanding of how we, as humans, fit into the scheme of earth. This unit suggests the areas where we can consider the planet not as ours to own but rather as a trust that we need to preserve as best we can.

After reading this unit you should be able to:

1. Describe the gaseous composition of the atmosphere.
2. Understand the role of carbon dioxide in the atmosphere.
3. Explain the nature of ozone depletion.
4. Distinguish the different layers of the atmosphere.

(B) KEY TERMS AND CONCEPTS

carbon dioxide cycle
climate
hydrologic cycle
ozone
photosynthesis
stratsosphere
troposphere
weather

(C) MULTIPLE CHOICE

_____ 1. Name the two major gases in the atmosphere. (p. 63)
(a) oxygen and carbon dioxide
(b) nitrogen and oxygen
(c) oxygen and water vapor

_____ 2. In the past 200 years, by what percent has the amount of carbon dioxide in the atmosphere risen? (p. 64)
(a) 25%
(b) 40%
(c) 10%

_____ 3. What function does the ozone perform? (p. 65)

(a) absorbs carbon dioxide
(b) absorbs ultraviolet radiation
(c) reflects carbon dioxide

_____ 4. What is the normal temperature lapse rate? (p. 66)

(a) 5^{0}C per km
(b) 2.5^{0}C per km
(c) 6.5^{0}C per km

_____ 5. When is ozone depletion most noticeable in Antarctica? (p. 68)

(a) early southern spring
(b) midsummer
(c) late southern winter

(D) TRUE OR FALSE

1. Carbon dioxide comprises only 0.04% of the air. (p. 64)
T _____ F _____

2. The warmer the air, the less water vapor it can hold. (p. 65)
T _____ F _____

3. Fair-weather skies look blue because more of the blue light is being scattered. (p. 65)
T _____ F _____

4. In the troposphere, temperatures either stay the same or increase with altitude. (p. 66)
T _____ F _____

5. The Montreal Protocol of 1987 was the first international effort to deal with ozone depletion. (p. 69)
T _____ F _____

(E) SHORT-ANSWER QUESTIONS

1. How does weather differ from climate? (p. 62)

____

____

2. Distinguish between the homosphere and the heterosphere. (p. 63)

3. Name the two main gases comprising the lower atmosphere and their respective percentages. (p. 63)

4. Name three examples of fossil fuels. (p. 63)

5. Describe the process of **photosynthesis**. (p. 64)

6. What role does carbon dioxide play in maintaining the earth's temperature? (p. 64)

7. Describe and explain the differences of water vapor content between the North Pole and the Equator. (p. 65)

8. What role do the oceans play in the **carbon cycle**? (p. 66)

9. How does ionization affect radio transmissions? (p. 70)

10. What is the **aurora borealis**? (p. 70)

____

(F) MATCHING

Match the following terms and their meanings:

_______ 1. oxidation		a. process whereby plants use carbon dioxide to form carbohydrates
_______ 2. constant gases		b. synthetic compounds that damage the ozone layer
_______ 3. climate		c. combination of oxygen and other materials to form new products
_______ 4. photosynthesis		d. layers in the atmosphere in which temperature increases with altitude
_______ 5. stratosphere		e. gases always found in the same proportion in the atmosphere
_______ 6. chloroflourocarbons		f. long-term conditions of aggregate weather
_______ 7. inversion		g. atmospheric layer of clear, calm air above the tropopause

(G) ESSAY QUESTIONS

1. How is carbon dioxide implicated in global climate change?

2. Describe the sequence of scientific findings that have brought ozone depletion to the attention of the world.

3. Why is knowledge of ultraviolet radiation important for the future?

(H) GRAPHICACY

1. Use the axis provided and draw a line graph for temperature and one for relative humidity as these undergo change during the course of a day. Use a solid line for temperature and a dashed line for relative humidity.

midnight noon midnight

2. Explain the relationship between temperature and relative humidity as illustrated in your completed sketch.

(I) HYPOTHESIS CONSTRUCTION

1. Write two statements that address the relationship between these two variables. The first statement uses temperature as the independent variable and the second uses relative humidity as the independent variable.

a.

b.

Unit 7

Radiation and the Heat Balance of the Atmosphere

(A) SUMMARY AND OBJECTIVES

Whether they are called shades or mirrors, sunglasses protect our eyes from those solar rays that seem to come at us from all different directions. While the sun may be always above us, its rays are reflected and re-reflected off a multitude of different surfaces. The amount of reflection (albedo) varies from place to place, creating significant temperature differences whether in a parked car on a summer's day or in the icy cold of Antarctica. The importance of understanding global temperatures is underscored by a special global warming section which examines the role of fossil fuel consumption in increasing carbon dioxide emissions. A computer projection for the year 2050 indicates the particularly severe impact of global warming on higher latitude areas.

After reading this Unit you should be able to:

1. Understand the principles of radiation.
2. Describe the scientific basis for global warming.
3. Understand the concept of heat balance.
4. Interpret the reasons for variations in earth's heat flows.

(B) KEY TERMS AND CONCEPTS

albedo	long wave radiation
conduction	net radiation
convection	radiation
greenhouse effect	sensible heat
latent heat	short wave radiation

(C) MULTIPLE CHOICE

____ 1.What percentage of all solar energy arriving at the outer edge of the atmosphere actually reaches the earth's surface? (p. 73)

(a) 53%
(b) 26%
(c) 72%

____ 2. By what percentage has the global carbon dioxide level increased in the last 50 years? (p. 75)

(a) 5%
(b) 10%
(c) 15%

____ 3. In which region does most latent heat loss occur? (p. 79)

(a) the desert
(b) the rainforest
(c) the mid-latitudes

____ 4. What probably caused the higher than normal warm years of the late 1980's. (p. 75)

(a) heating of the earth's core
(b) eruption of a volcano in the Philippines
(c) too many playoff games

____ 5. What is the term that describes the amount of heat energy required to raise the temperature of 1 gram of water by 10^0C? (p. 73)

(a) joule
(b) watt
(c) calorie

(D) TRUE OR FALSE

1. On March 22, a ray of solar energy falling on the equator is less likely to be reflected than one falling on 45^0N. (p. 73)
T_____ F_____

2. A cloudy winter is more likely to be warmer than a less cloudy one. (p. 74)
T_____ F_____

3. An asphalt surface has a higher albedo rate than a snow covered surface. (p. 73)
T_____ F_____

4. Gross radiation is the amount of radiation left over when all the incoming and outgoing radiation flows have been tallied. (p. 78)
T_____ F_____

5. The atmosphere is actually heated from below and not directly by the sun above. (p. 74)
T_____ F_____

(E) SHORT-ANSWER QUESTIONS

1. Distinguish between shortwave and longwave radiation. (p. 73)

2. What is the difference between solar and terrestrial radiation? (p. 74)

3. What is the function of **diffuse radiation**? (p. 73)

4. What three factors determine the amount of radiation that will be reflected by a surface? (p. 73)

5. Describe what is meant by sensible heat and mention one example of its occurrence. (p. 79)

6. How does the heat balance determine climate? (p. 79)

7. Define **ground heat flow**. (p. 79)

8. What is **latent heat of vaporization** and why is it significant for physical geographers? (p. 81)

____

____

9. What role do clouds play in determining radiant heat input? (p. 81)

____

____

10. Compare land and sea differences in **latent heat loss**. (p. 82)

____

____

(F) MATCHING

Match the following terms and their meanings:

_______ 1. radiation	a. the rate at which radiation is received on the outside of the atmosphere
_______ 2. sensible heat	b. transmission of energy in the form of electromagnetic waves
_______ 3. albedo	c. incoming solar radiation
_______ 4. soil heat flow	d. the temperature sensed by a person's body
_______ 5. solar constant	e. amount of incoming radiation that is reflected by a surface
_______ 6. conduction	f. heat conducted into and out of the earth's surface
_______ 7. insolation	g. transport of heat energy from one molecule to the next

(G) ESSAY QUESTIONS

1. How does albedo differ from the solar constant?

2. How has the Industrial Revolution affected weather and climate?

3. Describe the nature of the earth's "**greenhouse effect**".

(H) GRAPHICACY

1. Draw a sketch to illustrate the process of earth as a greenhouse.

2. Examine the heat balance graphs of Paris, France and Manaos, Brazil in Figure 7-6. Explain the following:

a. The differences in latent heat loss.

b. The differences in sensible heat loss.

c. The differences in net radiation.

3. Using Figure 7-6 in the text as a model, draw a comparable set of graphs for your own latitude and explain which one of the presented locations (Manaos, Aswan, Paris, Turukhansk) is most similar to your location.

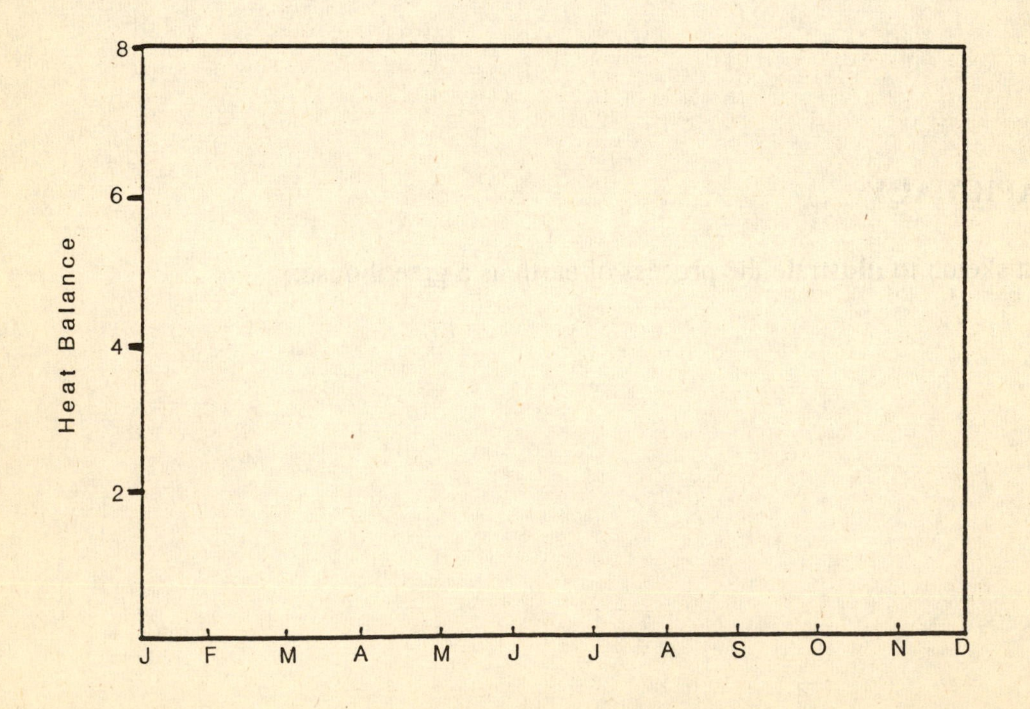

(I) HYPOTHESIS CONSTRUCTION

1. Write a hypothesis to describe the relationship between the color of the earth's surface and the amount of solar reflection.

Unit 8

Atmospheric and Surface Temperature

(A) SUMMARY AND OBJECTIVES

As we move towards building a comprehensive picture of weather characteristics and patterns, we need to isolate and understand each of the specific elements of weather. This unit presents temperature and temperature variations as the basis for our study of meteorology. The most important relationship that you should note in this section is between temperature and land and water bodies. The continental and maritime influences on temperature help explain why annual temperatures are so different in San Francisco and in Iowa City. In addition to naturally occurring temperature variations, there is also an increasing need to understand how human actions create changes in temperature patterns. Why do cities generally have more rain and more fog than rural areas? What causes smog in cities? These are questions that have to be addressed as our world becomes more urbanized.

After reading this unit you should be able to:

1. Understand the nature of heat and temperature.
2. Describe vertical temperature changes.
3. Explain how temperature is altered in urban areas.
4. Interpret maps of temperature variation.

(B) KEY TERMS AND CONCEPTS

adiabatic lapse	isotherms
advection	kinetic energy
continentality	net radiation
diurnal cycle	smog
dust domes	temperature inversion
environmental lapse rate	temperature gradient

(C) MULTIPLE CHOICE

____ 1. What happens to heat during condensation? (p. 86)

(a) it is given off
(b) it is absorbed
(c) it is replaced

____ 2. What is characteristic of unstable air? (p.86)

(a) flat, layered clouds
(b) cirrus clouds
(c) puffy, vertical clouds

_____ 3. Which place has the greatest annual temperature range? (p. 91)

(a) San Francisco
(b) Omaha
(c) Boston

____ 4) What is the temperature in degrees Celsius when we convert 80^0F? (p. 84)

(a) 21^0C
(b) 26^0C
(c) 30^0C

____ 5) What is the boiling point in the Kelvin scale? (p. 84)

(a) 100^0
(b) 373^0
(c) 212^0

(D) TRUE OR FALSE

1. The southern hemisphere has the greatest temperature range. (p. 92)
T_____ F _____

2. The dry adiabatic lapse rate is 10^0C per 1 km. (p. 85)
T_____ F_____

3) The horizontal movement of air is referred to as advection. (p. 92).
T_____ F_____

4) Isohyets are lines joining places of equal temperature. (p. 92)
T_____ F_____

5) Temperature inversion occurs only in valleys. (p. 87)
T_____ F_____

(E) SHORT-ANSWER QUESTIONS

1. Why is the Celsius scale preferred over the Fahrenheit scale in more than 90% of the earth's countries? (p. 84)

2. What vertical temperature change occurs within the troposphere? (p. 84).

3. What happens to heat during the process of **condensation**? (p. 86)

4. What are the two main factors associated with the spread of air pollution? (p. 88)

5. Explain why Los Angeles and Denver are among the worst U.S. cities in terms of air pollution. (p. 87-88)

6. What two factors reduce the amount of incoming short radiation in urban areas? (p. 88)

7. Why do the highest annual temperatures occur after the solstice? (p. 99)

8. Why does the Southern Hemisphere have smaller annual temperature ranges than the Northern Hemisphere? (p. 92)

9. What impacts do the ocean currents have on the temperatures in western Europe and in western South America? (p. 92)

10. Describe three reasons why cities have different albedo rates than rural areas. (p. 88)

____

____

(F) MATCHING

Match the following terms and their meanings:

_______	1. smog	a.	the energy of movement
_______	2. diurnal cycle of temperature	b.	horizontal movement of air
_______	3. isotherms	c.	horizontal rate of temperature change over distance
_______	4. advection	d.	pattern of temperature change during a day
_______	5. temperature gradient	e.	combination of chemical pollutants and fog
_______	6. kinetic energy	f.	cooling and expansion of rising air
_______	7. adiabatic lapse	g.	lines connecting places which have the same temperature

(G) ESSAY QUESTIONS

1. Explain how temperature differs from heat.

2. Explain why land and water have differing heating and cooling capabilities.

(H) GRAPHICACY

1. Examine Figure 8-11 and answer the following questions:

   a. Explain why the top map describes January conditions.

   b. Explain why the bottom map describes conditions in July.

2. Explain why the isotherms bend towards the equator over North America in Figure 8-11a.

3. Explain why the isotherms bend poleward over South America in Figure 8-11b.

4. After examining Table 8-1 and the unit opening photo, explain the temperature differences between the two hemispheres.

(I) HYPOTHESIS CONSTRUCTION

1. Complete this statement to explain the different temperature curves in Figure 8-9.

Places ____ have a greater temperature variation

than ____ because

____

2. Write a statement which shows the opposite relationship from the one in your sentence above.

Unit 9

Air Pressure and Winds

(A) SUMMARY AND OBJECTIVES

In the last few minutes of the football game, the wind changes direction and the outcome of the game is suddenly reversed! What caused the wind to blow from that direction? Why at this time in the afternoon? In this unit, we will see that such weather experiences are directly connected to the air pressure that exists in the atmosphere surrounding us. Here we examine air pressure and the forces that determine the direction and velocity of wind. We will determine the nature of wind patterns at various, related scales from the local beach level to the more global pattern that explains why most weather systems in the United States move from west to east. As you contrast air pressure to the other weather elements (temperature and humidity) you may be tempted to view air pressure as being "more equal than the others" in understanding our weather patterns.

After reading this unit you should be able to:

1. Understand the concept of air pressure.
2. Describe the relationship between wind and air pressure.
3. Relate air pressure variation to daily weather patterns.
4. Explain the causes of wind deflection.

(B) KEY TERMS AND CONCEPTS

air pressure gradient	isobar
anti-cyclone	katabatic wind
Chinook wind	land breeze
Coriolis force	Santa Ana
cyclone	sea breeze
geostrophic wind	wind chill factor

(C) MULTIPLE CHOICE

_____ 1. What percent of the planet's heat distribution is accounted for by atmospheric circulation? (p. 95)

(a) 55%
(b) 87%
(c) 75%

______ 2. How much more radiation does the equator receive when compared to the poles? (p. 97)

(a) 6 times
(b) 3 1/2 times
(c) 2 1/2 times

______ 3. What is standard sea-level air pressure in millibars? (p. 96)

(a) 998.4
(b) 1048.6
(c) 1013.2

______ 4. What percent of the atmosphere is found below 5 km? (p. 96)

(a) 25%
(b) 50%
(c) 75%

______ 5. Using the conversion factor of 33.865, what is 30.2" of air pressure equivalent to?

(a) 1022.7 mb
(b) 1015.6 mb
(c) 1124.3 mb

(D) TRUE OR FALSE

1. In the southern hemisphere, wind blows around an anticyclonic formation in a clockwise direction . (p. 102)
T_____ F_____

2. In the Northern Hemisphere wind is deflected to the right. (p. 98)
T _____ F _____

3. Between the equator and 35^0N outgoing radiation exceeds incoming radiation. (p. 97)
T_____ F _____

4. Coriolis force is greater at the poles than at the equator. (p. 98).
T_____ F _____

5. Track and field records stand a better chance of being broken at a stadium which is located at 2,240 meters than at sea level. (p. 100)
T _____ F _____

(E) SHORT-ANSWER QUESTIONS

1. Define **air pressure.** (p. 96)

2. Why does outgoing radiation exceed incoming radiation in the higher latitudes? (p. 97)

3. What two factors complicate the direct heat transfer to the poles? (p. 98)

4. Describe the nature of **Coriolis force.** (p. 98)

5. Describe the role of pressure-gradient force in wind formation. (p. 98)

6. Contrast the pressure-related characteristics of warm and cold air. (p. 98)

7. How does a **geostrophic** wind occur? (p. 99)

8. What is the difference between cyclonic and anticyclonic pressure systems? (p. 102)

____

____

9. What are meant by **katabatic winds**? (p. 102)

____

10. Describe the conditions associated with the Santa Ana wind. (p. 103)

____

____

(F) MATCHING

Match the following terms and their meanings:

_______	1. pressure gradient	a.	warm downwind on the eastern side of the Rockies
_______	2. mistral wind	b.	the area away from the wind, protected by topography
_______	3. Torricelli	c.	lines joining places of equal air pressure
_______	4. geostrophic	d.	difference in air pressure between two locations
_______	5. isobars	e.	cold, high velocity downwind in the Rhone Valley
_______	6. Chinook wind	f.	scientist who developed the mercury barometer
_______	7. leeward	g.	wind so deflected that it blows parallel to isobars

(G) ESSAY QUESTIONS

1. What two pieces of information are needed if we are to predict wind direction and wind velocity?

2. How does air pressure change with increasing altitude?.

(H) GRAPHICACY

1. Using Table 9-1, determine the windchill temperature in conditions where the temperature is 5^0F and the wind velocity is 25 m.p.h.

____

2. Draw a sketch of a cyclonic system in the northern hemisphere, indicating wind direction and isobar values (use an isobar interval of 4 mb, with a low of 996 mb and the highest value being 1012 mb).

3. Study Figure 9-9 and answer the following questions:

a. What is the isobar interval?

b. In which hemisphere is this wind located?

(I) HYPOTHESIS CONSTRUCTION

1. Examine Figure 9-3 and explain what is happening at the 40^0 latitude.

2. Write a hypothesis statement which addresses this change.

Unit 10

Circulation Patterns of the Atmosphere

(A) SUMMARY AND OBJECTIVES

As with most things in nature, winds are not mere random phenomena. Winds blow from certain directions in accordance with particular principles. When we speak of a westerly wind, we are not just referring to the direction from which the wind blows, but also to a set of weather patterns which can then be expected to follow. In this section, we examine those wind patterns that provide some predictability to our weather. The most prominent feature of this section is that wind is indeed very strongly related to the existing pressure systems detailed in Unit 9. Air pressure (and its associated wind) is therefore the most dynamic factor of the "sea of air" in which we live.

After studying this unit, you should be able to:

1. Understand the principles of global atmospheric circulation.
2. Identify the major global air pressure systems.
3. Discuss the differential impact of land and sea on pressure systems.
4. Explain upper air circulation patterns.

(B) KEY TERMS AND CONCEPTS

azonal flow	Polar Easterlies
ITCZ	ridges
jet stream	trade winds
monsoons	troughs
polar highs	zonal flow

(C) MULTIPLE CHOICE

____ 1. From which direction do the southern hemisphere trade winds blow? (p. 107)

(a) northeast
(b) southeast
(c) northwest

____ 2. Between which latitudes are the Westerlies found ? (p. 107)
(a) 30 to 40
(b) 10 to 30
(c) 30 to 60

____ 3. Between which two latitudes does the annual maximum solar heating shift? (p. 108)
(a) 23 1/2^{0}N and 47^{0}N
(b) 23 1/2^{0}S and 66 1/2^{0}N
(c) 23 1/2^{0}N and 23 1/2^{0}S

____ 4. At what latitude is the Equatorial Low situated in July? (p. 108)
(a) 15^{0}N
(b) 25^{0}N
(c) 25^{0}S

____ 5. In monsoon India, which month receives the most rainfall? (p. 110)
(a) January
(b) March
(c) July

(D) TRUE OR FALSE

1. The Polar winds blow from west to east. (p. 107)
T ____ F ____

2. Along the Polar Front, the warmer subtropical air is forced to rise over the colder and denser polar air. (p. 107)
T ____ F ____

3. The Subtropical Highs are most evident during the winter. (p. 109)
T ____ F _____

4. The dominant high pressure cell over northern Asia is called the Siberian High. (p. 110)
T _____ F _____

5. Meridional air movement refers to the north-south exchange of air. (p. 112)
T _____ F _____

(E) SHORT-ANSWER QUESTIONS

1. What do we call the belt of rising air in the equatorial region? (p. 106)

___

2. Why do we refer to some winds as **trade winds**? (p 108)

___

3. In what way do the circulation patterns of Northern and Southern Hemispheres differ? (p. 106)

___

4. What effect does continentality have on air pressure system formation? (p. 108)

___

5. Name the two separate cells of the Northern Hemisphere Polar High. (p. 109)

___

6. Explain why the northern latitudinal pressure systems weaken during July. (p. 108)

___

7. Why do the Pacific Highs have less of a north-south shift than the Equatorial Low? (p. 109)

___

___

8. How does **azonal flow** influence temperatures in:
   a. the polar regions (p. 113)

   ___

   b. in the tropics (p. 113)

   ___

9. Which area in the northern latitudes has the most extreme seasonal cooling? Why? (p. 110)

___

10. What distinguishes the subtropical jet stream from the Polar Front jet stream? (p. 101)

____

____

(F) MATCHING

Match the following terms and their meanings:

_______	1. Aleutian Low	a. zone of calm at the equator
_______	2. azonal flow	b. regional circulation systems
_______	3. Polar Front	c. westward flowing winds
_______	4. doldrums	d. air pressure system in the northeastern Pacific
_______	5. monsoons	e. nonwesterly flow of air
_______	6. trade winds	f. sharp atmospheric boundary line in the higher latitudes
_______	7. secondary circulation	g. seasonal reversal of winds, especially in Asia

(G) ESSAY QUESTIONS

1. Describe the two factors which make the global circulation model much more complex than would otherwise be the case.

2. Compare and contrast the migration patterns of air pressure patterns in the two hemispheres.

(H) GRAPHICACY

1. Complete and label the sketches to show the monsoons in India; indicate the ITCZ, air pressure systems, and wind direction.

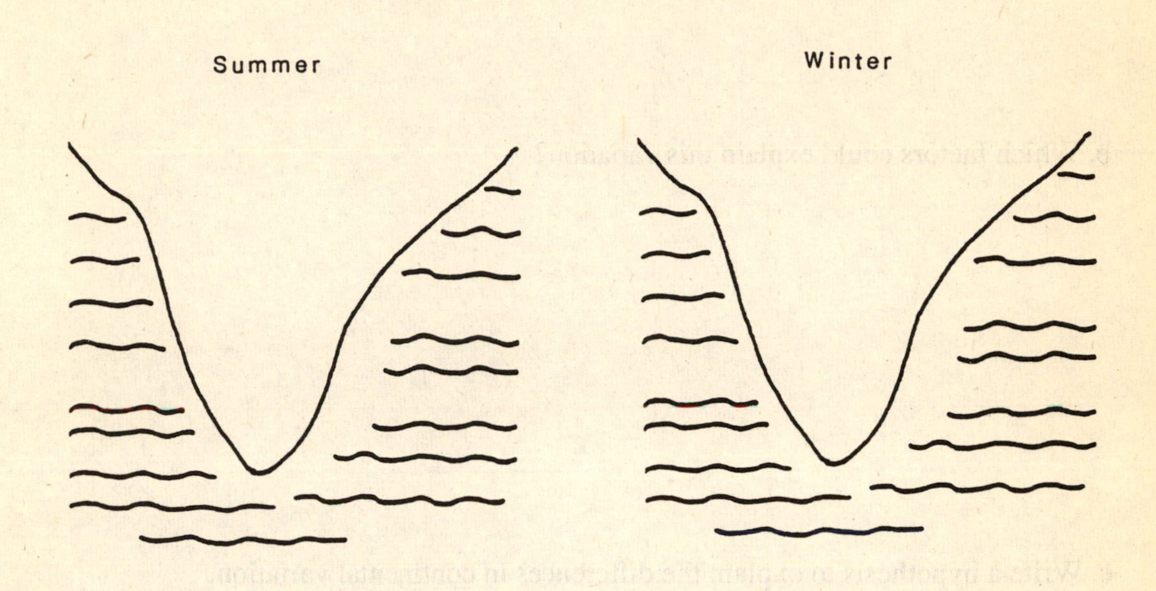

2. On the July map (Figure 10-3), trace the latitudinal line of your college town. Explain the nature of the isobar fluctuations where they intersect your latitude line.

(I) HYPOTHESIS CONSTRUCTION

1. Examine Figure 10-3 and answer the questions:

   a. Which continent has the greatest seasonal air pressure variation?

   b. Which factors could explain this variation?

   c. Write a hypothesis to explain the differences in continental variation.

Unit 11

Hydrosphere: Circulation of the World Ocean

(A) SUMMARY AND OBJECTIVE

The continuing international interest in the America's Cup boat race could be seen as an attempt by humans to attempt to conquer another medium - the ocean. But while many of us may dream about sailing solo around the world, when we examine a globe we are soon daunted by the enormity of this vast area of world water. In this unit we explore those world oceans with all of their movement and changeability. In addition to what happens at the water level, we soon discover that this hydrosphere has enormous impacts on the weather and climate conditions in our atmosphere, a fact that finds its most dramatic expression in the El Nino phenomenon of ocean warming. So whoever wants to win the America's Cup, they better be sure that they know the geography of those oceans.

After studying this unit you should be able to:

1. Explain the relationship between ocean and atmospheric circulation.
2. Understand the flow behavior of ocean currents.
3. Identify the major ocean currents.
4. Determine the role of ocean currents in maintaining global heat balance.

(B) KEY TERMS AND CONCEPTS

drift	ocean currents
El Nino	Southern Oscillation
ENSO	thermohaline
gyres	upwelling

(C) MULTIPLE CHOICE

_____ 1. What percentage of the earth is covered by ocean? (p. 114)
(a) 57%
(b) 62%
(c) 71%

_____ 2. The oceans account for what percentage of the total movement of heat from low to high latitudes? (p. 114)

(a) 10%
(b) 13%
(c) 52%

_____ 3. What is the average speed of ocean currents? (p. 115)

(a) 55 miles per hour
(b) 25 miles per hour
(c) 5 miles per hour

_____ 4. In what direction do ocean currents in the Northern Hemisphere get deflected? (p. 116)

(a) to the west
(b) to the right
(c) to the left

______ 5. Name an important benefit of upwelling. (p. 119)

(a) It brings warmer temperatures.
(b) It brings fish nutrients.
(c) It decreases coastal erosion.

(D) TRUE OR FALSE

1. The ocean beneath a sub-tropical region is more saline than under an equatorial belt. (p. 115)
T _____ F _____

2. Gyres refer only to the clockwise water circulations. (p. 116)
T _____ F _____

3. The center of a gyre is normally associated with calm weather. (p. 117)
T _____ F _____

4. The sub-polar gyres are not present in the southern hemsiphere because of the absence of large continents. (p. 118)
T _____ F______

5. Some of the moister coastal areas are associated with upwelling. (p. 119)
T _____ F _____

(E) SHORT-ANSWER QUESTIONS

1. What function do the oceans and the atmospheric circulation have in common? (p. 114)

____

2. Why do warm ocean currents give up more moisture to the atmosphere? (p. 115)

____

3. Why are ocean currents also called **drifts**? (p. 115)

____

4. Where are the fastest moving currents found? (p. 115)

____

5. What is the leading generator of ocean currents? (p. 115)

____

6. At what depth does wind-generated motion cease? (p. 116)

____

7. How do **gyres** differ from eddies? (p. 116)

____

____

8. Which areas are most associated with upwelling? (p. 119)

____

9. Contrast the heat storage capacity of land and ocean areas. (p. 124)

____

____

10. Describe the two ways in which ocean evaporation influences climate. (p. 124)

____

____

(F) MATCHING

Match the following concepts and their meanings:

_______	1. upwelling	a. cold current off southwest coast of Africa
_______	2. El Nino	b. large, circular ocean current loops
_______	3. Kuroshio	c. upper movement of colder, subsurface water
_______	4. thermohaline	d. large scale fluctuation of air pressure in the South Pacific
_______	5. gyre	e. anomalous warming of ocean off the coast of South America
_______	6. Benguela	f. warm current off the coast of Japan
_______	7. ENSO	g. deep-sea, temperature dependent oceanic movement

(G) ESSAY QUESTIONS

1. Explain why the piling up of ocean water occurs more on the western side than the eastern side of oceans.

2. Why is the ocean more saline in the subtropical high pressure zone than in the equatorial belt?

3. Name two reasons why deep-sea currents are much slower than their surface counterparts.

4. Describe the nature of countercurrents.

5. What role does the Bering Strait play in ocean current movements?

6. Describe the scientific findings and theory-building surrounding the El Nino phenomenon.

(H) GRAPHICACY

1. Compare Figure 11-5 to the Atlantic Ocean section of Figure 10-3. How does one of these illustrations help us explain the other?

2. Refer to the unit opening photo on page 103 of the text. Explain what is happening in the northern part of Africa and off the west coast of South America.

(I) HYPOTHESIS CONSTRUCTION

1.The El Nino phenomenon has many scientists searching for explanations. List three possible causes for this climatic phenomena.

2. Construct a hypothesis that you think best addresses El Nino.

Unit 12

Atmospheric Moisture and the Water Budget

(A) SUMMARY AND OBJECTIVES

The water that we consume after an energetic sports activity and the water locked into the Antarctic glaciers are both included in our larger water system referred to as the hydrosphere. Apart from its role as a life-giver and a source of summer sport, water does give us much of the weather and climate that we experience. Here we examine our planet's H_20 from a geographic perspective - this means we want to find out where that water is stored and used and what the consequences are of these patterns. We will explore those forms of water that we see - the clouds and precipitation - as well as the water in the soil and contained within our plants.

After studying this unit you should be able to:

1. Identify the changes in water from one form to another.
2. Understand the role of relative humidity in atmospheric change.
3. Classify cloud types.
4. Explain differences in the water balance across the surface of the earth.

(B) KEY TERMS AND CONCEPTS

calorie
coalescence
condensation nuclei
dew point
evapotranspiration
hydrosphere
latent heat of fusion
precipitation
relative humidity
saturation
sublimation
vaporization

(C) MULTIPLE CHOICE

_____ 1. What term refers to the ratio of the mass of water vapor to the total mass of the dry air containing the water vapor? (p. 127)

(a) relative humidity
(b) mixing ratio
(c) specific humidity

_____ 2. In the United States, what percentage of national water withdrawal is taken by agriculture? (p. 130)

(a) 25%
(b) 40%
(c) 70%

_____ 3. What name is given to stratus clouds that have precipitation occurring? (p. 131)

(a) cirrostratus
(b) vapostratus
(c) nimbostratus

_____ 4. At which latitude does the greatest evaporation occur? (p. 134)

(a) 0^0
(b) 25^0
(c) 55^0

_____ 5. What percent of the world's water is in freshwater form? (p. 128)

(a) 3%
(b) 15%
(c) 50%

(D) TRUE OR FALSE

1. Cirrus clouds occur mostly at altitudes above 6 km. (p. 132)
T _____ F _____

2. Sublimation refers to the temperature at which all condensation occurs. (p. 126)
T _____ F _____

3. The relative humidity is usually lowest when the temperature is at its highest. (p. 127)
T _____ F _____

4. Glaze refers to pellets of ice produced by the freezing of rain before it reaches the ground. (p. 132)
T _____ F _____

5. Rates of precipitation and runoff are highest near the equator. (p. 134)
T _____ F _____

(E) SHORT-ANSWER QUESTIONS

1. Describe the latent heat of fusion. (p. 126)

2. Define **relative humidity**. (p. 127)

3. What is meant by **saturated air**? (p. 126)

4. Why are condensation nuclei important for weather change? (p. 131)

5. What are the three criteria for classifying clouds? (p. 131)

6. Describe the characteristics of cirrus clouds. (p. 132)

7. What weather is usually associated with cumulonimbus clouds? (p. 131)

8. How are hailstones formed? (p. 132)

9. What are the two factors that determine the **water balance**? (p. 132)

10. What type of weather is normally produced by permanent high pressure cells? (p. 135)

____

(F) MATCHING

Match the following concepts and their meanings:

_______ 1. sublimation
_______ 2. psychrometer
_______ 3. hydrosphere
_______ 4. runoff
_______ 5. calorie
_______ 6. transpiration
_______ 7. advection

a. passage of water through leaf pores
b. removal of water by overland or stream flow
c. horizontal movement of air
b. water vapor changing directly into ice
e. instrument to measure relative humidity
f. the global water system
g. the amount of energy required to raise the temperature of 1 g of water 1^0C

(G) ESSAY QUESTIONS

1. Describe the two conditions necessary for evaporation to occur.

2. Contrast the Bergeron and the coalescence processes.

(H) GRAPHICACY

1. The following is an excerpted humidity table:

Depression, ^{0}F >>

Air Temp ^{0}F	5	6	7	8	9
50	61	55	49	43	38
55	70	65	59	54	49
60	68	63	58	53	48
65	75	70	66	61	56
70	77	72	68	64	59

a. If the dry bulb reading is 60^{0}F and the wet bulb reading is 54^{0}F, what is the depression?

b. Now determine the relative humidity for the values in 1(a).

2. Draw a sketch of the hydrologic cycle indicating the major elements of transfer.

3. Examine Figure 12-10.

a. Draw a cross-section through the global precipitation map at the 25°E longitude, using the axes provided.

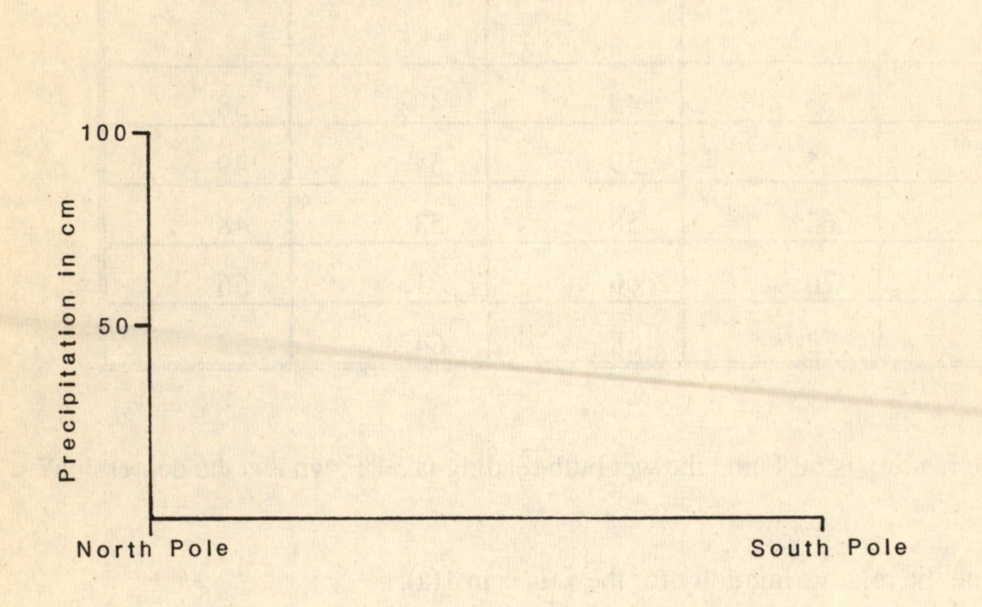

b. Describe and explain the pattern of precipitation that emerges in your sketch.

(I) HYPOTHESIS CONSTRUCTION

1. Examine Figure 12-9.

   a. Describe the differences in precipitation and evaporation at 50^0N and 50^0S.

   b. Construct a hypothesis to explain the contrast.

Unit 13

Precipitation, Air Masses, and Fronts

(A) SUMMARY AND OBJECTIVES

Wherever, we are, we do know that the weather never stays the same for very long. This dynamic aspect of the atmosphere is often caused by the movement of large bodies of air which move from one area to the next. In this unit, we learn just why a summer's day at the beach and a mid-winter day are so different in temperature and precipitation. The study of air masses is central if we are to be able to predict the type of weather that will be present on graduation day or for a major sports event. While our weather predictions for a picnic day may not be life altering, other weather phenomena such as tornadoes do represent a threat to human life. In our study of the geography of tornadoes we soon discover that the Kansas setting for Dorothy and Toto was (and still is) correct. The existence of such a "Tornado Alley" can be explained, if not yet controlled.

After reading this unit you should be able to:

1. Describe the three types of precipitation.
2. Interpret precipitation maps.
3. Explain the weather associated with warm and cold fronts.
4. Identify the features of the main air masses in North America.

(B) KEY TERMS AND CONCEPTS

adiabatic cooling	Doppler radar
air mass	front
Chinook winds	ITCZ
convection	orographic precipitation

(C) MULTIPLE CHOICE

_____ 1. At what latitude is the ITCZ located in July? (p. 138)

(a) 5^0N
(b) 15^0S
(c) 10^0N

_____ 2. At what rate does the dry adiabatic lapse process occur? (p. 139)

(a) 10^0C per km

(b) 15^0C per km

(c) 20^0C per km

_____ 3. By what name are lee-side, warm, dry winds known? (p. 144)

(a) Cascades

(b) Mushings

(c) Foehn

_____ 4. What type of precipitation is most common in the equatorial zone? (p. 141)

(a) cyclonic

(b) frontal

(c) convectional

_____ 5. What is the rate for Saturated Adiabatic Lapse? (p. 142)

(a) 4.0^0 per 1km

(b) 6.5^0 per 1km

(c) 8.0^0 per 1km

(D) TRUE OR FALSE

1. The National Storms Forecast Center is located in Kansas. (p. 142)
T _____ F _____

2. The mP air mass originates in the Gulf of Mexico. (p. 146)
T _____ F _____

3. Cyclonic precipitation occurs in hurricanes. (p. 145)
T _____ F _____

4. Warm fronts are associated with cumulonimbus clouds. (p. 145)
T _____ F _____

5. Thunderstorms actually consist of several clustered cells. (p. 140)
T _____ F _____

(E) SHORT-ANSWER QUESTIONS

1. Name the four processes that are needed to produce precipitation. (p. 138)

___

2. Why is air instability necessary in cumulonimbus clouds? (p. 139)

___

___

3. What are the two main factors determining the location of the **ITCZ ?** (p. 138)

___

4. How do **squall line** storms affect weather? (p. 139)

___

___

5. How do the mechanisms for thunder and lightning diffe?. (p. 140)

___

___

6. Why do most convectional storms occur in the tropics? (p. 141)

___

7. Why do converging warm and cold air masses not mix easily? (p. 144)

___

8. What are the three factors distinguishing one **air mass** from another? (p. 145)

___

___

9. What role does the angle of the frontal slope have on precipitation? (p. 145)

___

___

10. Name the source regions of cT and cA air masses. (p. 146)

___

___

(F) MATCHING

Match the following terms and their meanings:

_______	1. squall line	a. small vortex of rapidly rotating air
_______	2. convection	b. clusters of thunderstorm cells
_______	3. cP air mass	c. air mass originating in the tropical, ocean area
_______	4. supercells	d. a downslope dry and warm wind
_______	5. tornado	e. air mass originating in polar, continental areas
_______	6. foehn	f. area ahead of the front with intense thunderstorms
_______	7. mT air mass	g. vertical movement of heat

(G) ESSAY QUESTIONS

1. Describe the two conditions necessary for cumulonimbus clouds to develop.

2) Describe the features of the three stages of thunderstorm development.

a. ______________ stage

b. ______________ stage

c. ______________ stage

3. The movie *The Wizard of Oz* has long been a favorite of many.

a. Where were Dorothy and Toto when the tornado occurred.

b. Explain why this part of the country was (and still is) prime area for tornado occurrence.

(H) GRAPHICACY

1. Describe, with the aid of three sketches, what happens when a cold front passes over a city.

a. Time 1 ... cloud formation

b. Time 2 ... precipitation

c. Time 3 ... clearing skies

(I) HYPOTHESIS CONSTRUCTION

1. Compare the precipitation map in Figure 13-8 with a physical relief map in an atlas and answer the following questions.

a. Locate and identify the Coastal Range and the Cascades.

b. Why does the precipitation increase to 100 cm on the northeastern border of Washington state.

c. Construct a hypothesis to explain this general principle.

Weather Systems

(A) SUMMARY AND OBJECTIVES

When we watch the weather report on television, we are seeing a simulated movement of weather systems as they move from one part of the atmosphere to another. This dynamic or action-oriented aspect of weather reminds us that the weather system is never stationary or boring. The two most significant systemic movements that affect most of the world's populations are midlatitude cyclones (in the zone 20^0 to 50^0 north and south) and hurricanes (originating in the tropics but often affecting regions as far away as 45^0 from the equator). After considering the conditions that are necessary for the formation of hurricanes, the authors reflect on their occurrence and the death and damage caused by this powerful atmospheric force. More benign are the low pressure systems in the midlatitude zone (35^0 to 50^0 from the equator). While muted in its strength, temperate cyclones are vitally important in the mid-to-higher latitudes because they bring the penetrating rainfall so essential to agriculture in those heavily populated regions of the world.

After studying this unit you should be able to:

1. Explain the conditions under which hurricanes develop.
2. Understand the weather forces operating in the midlatitudes.
3. Describe the nature of midlatitude cyclones.
4. Explain the interactions in weather in systems terms.

(B) KEY TERMS AND CONCEPTS

cyclogenesis	jet stream
easterly wave	occluded front
eye	stationary front
hurricane	typhoons

(C) MULTIPLE CHOICE

_____ 1. At what sustained wind speed does a tropical depression become a tropical storm? (p. 149)

(a) 25 mph
(b) 34 mph
(c) 44 mph

_____ 2. What is the name given to hurricanes in the Indian Ocean? (p. 150)

(a) cyclones
(b) typhoons
(c) spouts

_____ 3. When does the annual North atlantic hurricane season officially start? (p. 151)

(a) May 24
(b) January 18
(c) June 1

_____ 4. Name the major, destructive hurricane which struck Miami and the Gulf of Mexico in 1992. (p. 152)

(a) Nancy
(b) Andrew
(c) Daphne

_____ 5. At what approximate height does the jet stream travel? (p.154)

(a) 10,000 ft
(b) 20,000 ft
(c) 30,000 ft

(D) TRUE OR FALSE

1. Temperatures vary a great deal throughout the area of a tropical cyclone. (p. 150)
T_____ F_____

2. Hurricanes occur primarily during the late summer and autumn. (p. 151)
T _____ F_____

3. The polar front jet stream circles the globe in a path parallel to the lines of latitude. (p. 154)
T _____ F_____

4. Midlatitude cyclones generally move from west to east. (p. 154)
T _____ F_____

5. Temperate cyclones provide the greatest source of rain in the midlatitude zones. (p. 156)
T _____ F_____

(E) SHORT-ANSWER QUESTIONS

1. What function does the **easterly wave** play in tropical storm development?. (p. 149)

____

____

2. What two factors determine the intensity of a hurricane? (p. 150)

____

3. Between which latitudes do hurricanes originate? (p. 151)

____

4. Why are hurricanes not found within 5^0 of the equator? (p. 151)

____

5. What happens to the strength of a hurricane when it passes over a landmass? (p. 152)

____

6. What is the major difference between circulation in the tropics and in the higher latitudes? (p. 153)

____

7. What does the **polar front** separate? (p. 154)

____

8. What is the shape of the isobar pattern in an open-wave cyclone? (p. 155)

____

9. Why are the midlatitude cyclones important for agriculture? (p. 156)

____

10. What are the two major inputs into the weather system? (p. 157)

____

(F) MATCHING

Match the following terms and their meanings:

_______	1. cyclogenesis	a.	names given to hurricanes in the western Pacific
_______	2. willy-willies	b.	dramatic rise in sealevel caused by a storm
_______	3. eye	c.	a front created when a cold front overtakes a warm front
_______	4. storm surge	d.	formation and movement of midlatitude cyclones
_______	5. typhoons	e.	weak low pressure tropical systems
_______	6. occlusion	f.	name given to hurricanes in Australia
_______	7. tropical depression	g.	the central area of calmness in a hurricane

(G) ESSAY QUESTIONS

1. What are the differences between a tropical depression and a tropical storm?

2. Describe the role of heat transfer in the creation of hurricanes.

3. Both tropical and midlatitude cyclones are firstly characterized by having low pressure centers. But there are also other major differences. Compare and contrast these different characteristics in table form.

	Tropical Cyclones	Midlatitude Cyclones
Location		
Air Pressure		
Direction		
Wind		
Speed		
Precipitation		
Season		

(H) GRAPHICACY

1. Figure 14-2 is a view of Hurricane Gilbert. How could we determine from this image that the hurricane is occurring in the northern hemisphere?

2. Describe the midlatitude cyclone in Figure 14-10a in system terms.

(I) HYPOTHESIS CONSTRUCTION

1. List three factors that influence the severity of hurricanes.

2. Which of these will be the most prominent in the next 50 years?

3. Construct a hypothesis that a geographer in the field can use to determine future hurricane severity.

Unit 15

Weather Tracking and Forecasting

(A) SUMMARY AND OBJECTIVES

And now, you finally get your chance to be the weather forecaster! The National Weather Service synoptic maps are not only fun to look at but they also contain most of the information that we need to start our exercise in weather forecasting. After we stop marveling at the organizational feat that brings all of this data together in one place and on one map, we can appreciate how this map tool can help make our lives a little better and a little bit more predictable. We are also told that increasingly, private weather forecasting companies provide a variety of weather data services for individualized corporate needs. Good luck in your weather prediction!

After reading this unit you should be able to:

1. Explain the process of collecting national weather data.
2. Understand the role of satellites in weather analysis.
3. Interpret a synoptic weather map.
4. Appreciate the complexity involved in weather forecasting.

(B) KEY TERMS AND CONCEPTS

500-mb chart
geosynchronous orbit
radiosonde
rawinsonde
storm track
synoptic weather map

(C) MULTIPLE CHOICE

_____ 1. In which federal department is the Weather Bureau housed? (p. 160)
(a) Agriculture
(b) Defense
(c) Commerce

_____ 2. Over which location does the GOES-9 satellite hover? (p. 161)
(a) eastern Atlantic
(b) eastern Pacific
(c) western Pacific

_____3. What is the average altitude of weather satellites? (p. 161)
(a) 200 miles
(b) 400 miles
(c) 700 miles

_____ 4. What is the difference in time between two consecutive X symbols on a storm track? (p. 164)
(a) 3 hours
(b) 6 hours
(c) 10 hours

_____ 5. Where is the U.S. National Severe Storms Forecast Center located? (p. 170)
(a) Kansas City
(b) Miami
(c) Bowling Green

(D) TRUE OR FALSE

1. Satellites in geosynchronous orbit are stationary above a given location. (p. 161)
T _____ F _____

2. Wind direction is always indicated by the direction from which the wind is blowing. (p. 164)
T _____ F _____

3. The higher the altitude of the 500-mb level, the colder the surface temperature of the air beneath it. (p. 164)
T _____ F _____

4. Winds on the 500-mb chart flow parallel to the isobars. (p. 164)
T _____ F _____

5. The world's highest air pressure has been recorded in Brazil. (p. 169)
T _____ F _____

(E) SHORT-ANSWER QUESTIONS

1. What time system is used to determine when to take global weather observations? (p. 160)

____

2. List the types of data reported to the National Weather Service. (p. 160)

___

___

3. How often are higher airborne weather readings transmitted? (p. 161)

___

4. Explain what is meant by **GOES**. (p. 161)

___

5. What will the geographic coordinates of the GOES-8 satellite be? (p. 161)

___

6. What is the main publication of the National Weather Service? (p. 162)

___

7. How does the altitude of the **500-mb level** relate to surface temperature? (p. 164)

___

___

8. What does the triangular pennant on the 500-mb map indicate? (p. 167)

___

9. Describe a 500-mb trough. (p. 167)

___

___

10. What is the main private sector client-group for weather information? (p. 171)

___

(F) MATCHING

Match the following terms and their meanings:

_______ 1. synoptic weather
_______ 2. rawinsonde
_______ 3. radiosonde

a. radio-equipped instruments carried aloft
b. summarized weather information
c. radar trackings of air-borne instruments

(G) ESSAY QUESTIONS

1. Explain why the main Daily Weather Map describes the *7:00 a.m.* Eastern Standard Time conditions. Hint: consider the global dimensions of weather forecasting.

2. Describe the role of the 500-mb chart in weather forecasting.

(H) GRAPHICACY

1. Describe the weather conditions at Amarillo, Texas (Fig. 15-4).

Temperature	__________
Dew point	__________
Air pressure	__________
Pressure tendency	__________
Wind direction	__________
Wind velocity	__________
Cloud cover	__________

2. Refer to Figure 15-4.
   a. Which station on the map has the highest temperature?

   b. What is the highest temperature?

   c. What is the lowest isobar value?

3. Refer to Figure 15-4.

   a. Using the linear scale at the bottom left of the map, calculate the distance traveled by the midlatitude cyclone.

   b. Calculate the speed (in mph) of the cyclonic system.

c. Describe the weather changes we can expect in Cleveland within the next 24 hours.

(I) HYPOTHESIS CONSTRUCTION

1. Figure 15-9 shows global weather extremes.

   a. Identify the location of the Northern Hemisphere's lowest temperature.

   b. Why does this place have lower temperatures than places further north?

   c. Refer back to Unit 8 and construct a hypothesis to explain the location of the lowest temperature.

Unit 16

Climate Classification and Regionalization

(A) SUMMARY AND OBJECTIVES

The urge to classify things (socks by color, books by size) overtakes all of us at some point. The scientific classification of phenomena is necessary if we are to understand the patterns of the objects we are studying. Climate is especially suitable for classification because it is global in scope and has clearly determined features. While there are many different climate classification schemes, we focus here on the most well-known one, the Koppen system. This system is direct and logical and it will form the framework for Units 16-18 that follow. Remember it is the logic of the system that must first be grasped in the hypothetical model before it can be applied to a real world setting. Go forth and classify!

After studying this unit you should be able to:

1. Understand the objectives of climate classification.
2. Explain the basis for the Koppen classification.
3. Describe the climate types in the Koppen classification.
4. Evaluate the strengths and weaknesses of the Koppen system.

(B) KEY TERMS AND CONCEPTS

climate
mesothermal
microthermal
monsoon
savanna
steppe
temperate
tundra

(C) MULTIPLE CHOICE

_____ 1. What is the minimum averaging period for climate data? (p. 175)
(a) 20 years
(b) 30 years
(c) 50 years

_____ 2. What was the key for Koppen in his climate classification system? (p. 175)
(a) precipitation
(b) temperature
(c) vegetation

_____ 3. On which coasts are the Cs climates found? (p. 179)
(a) west
(b) south
(c) east

_____ 4. Where is the Gobi desert located? (p. 179)
(a) India
(b) China
(c) South Africa

_____ 5. What is the climate type in Miami? (p. 182)
(a) savanna
(b) tropical rainforest
(c) Cf

(D) TRUE OR FALSE

1. The microthermal climate has at least one month greater than 10^0C. (p. 176)
T _____ F _____

2. The minimum monthly temperature for the Af climate is 50^0F. (p. 177)
T _____ F _____

3. Steppe vegetation is characterized by tall trees and thorny bushes. (p. 179)
T _____ F _____

4. In the U.S. the Cf climates zone is extended by the influence of the warm Gulf of Mexico. (p. 180)
T _____ F _____

5. The boundaries of climate zones can change over a 40-year period. (p. 183)
T _____ F _____

(E) SHORT-ANSWER QUESTIONS

1. How does climatology differ from meteorology? (p. 174)

____

____

2. Why are lines on a climate map somewhat misleading? (p. 175)

____

3. What is the main feature of the *Aw* climate? (p. 178)

____

____

4. Why does the A climate not extend far north in Africa? (p. 179)

____

5. What climates are associated with the subtropical high pressure zones? (p. 179)

____

6. How does the BW climate differ from the BS? (p. 179)

____

7. What does the change in the third **Koppen** letter from k to h mean? (p. 179)

____

8. Which Southern Hemispheric landmass best illustrates the pattern of B climates? Explain why. (p. 179)

____

____

9. What is responsible for the interruption of the east-west belt of C climates in the Northern Hemisphere? (p. 179)

____

10. What are the temperature requirements that distinguish the ET from the EF climates? (p. 182)

____

____

(F) MATCHING

Match the following terms and their meanings:

_______	1. tundra	a. an A climate with a dry winter season
_______	2. climate	b. humid climates with moist winters
_______	3. mesothermal	c. cold climates of the high latitudes
_______	4. B climates	d. transitional climate between BW and moister climates
_______	5. savanna	e. having a moderate amount of heat
_______	6. Cf climates	f. average values of weather elements over a 30-year period
_______	7. steppe	g. desert climates

(G) ESSAY QUESTIONS

1. List the five objectives of an ideal climate classification system.

2. Explain why Eurasia is the only continent containing Dw climates.

(H) GRAPHICACY

1. Answer the following questions based on Figure 16-2.

   a. Why is the Af climate larger on the east coast of the hypothetical continent?

   b. Why does the BW climate in the center of the continent extend poleward in both hemispheres?

   c. Why does the Aw not extend to the east coast?

(I) HYPOTHESIS CONSTRUCTION

1. Examine Figure 16-4 and answer the following questions.

   a) Use an atlas to determine why the 1941 line makes a southerly dip in the central part of the peninsula.

b. Speculate as to why the 64.4^{0}F line in Florida has shifted southward in this 38-year period.

c. Construct a hypothesis that explains this type of isotherm movement.

Unit 17

Tropical (A) and Arid (B) Climates

(A) SUMMARY AND OBJECTIVES

By this time in the course you have probably heard it said many times that geography is concerned with the interactions of humans and the environment. This unit focuses on such interaction by examining natural landscapes (rainforests for example) that are being altered by human action. The climate zones that we examine do of course change naturally, but change occurs over a long span of time. Our concern as geographers studying deforestation and the spreading of deserts, is that change is occurring in a much shorter time frame than at which we or the planet can adapt. If we are to be stewards of this planet, then we need to arm ourselves with the basic facts and use our knowledge and skills to take better care of Mother Earth. It's the only planet we have.

After reading this unit, you should be able to:

1. Understand the distinctions within the A and B climates.
2. Interpret climograms of tropical and arid climates.
3. Evaluate the impact of human activity on fragile earth zones.
4. Explain the extent and consequences of deforestation and desertification.

(B) KEY TERMS AND CONCEPTS

annual march of temperature
climograph
deforestation
desertification
fertigation
Sahel

(C) MULTIPLE CHOICE

_____ 1. What is the minimum monthly average temperature for the A climates? (p. 186)
(a) 55°F
(b) 60°F
(c) 64°F

_____ 2. What type of precipitation is typical of the A climates? (p. 186)
(a) convectional
(b) orographic
(c) cyclonic

_____ 3. Which ocean current causes the fog off the coast of Namibia? (p. 190)

(a) Humboldt
(b) Miller
(c) Benguela

_____ 4. By what other name is the steppe in North America known? (p. 191)

(a) long-grass prairie
(b) short-grass prairie
(c) dry desert

_____ 5. What percent of the planet's species are found in the tropical rainforest. (p. 185)

(a) 50%
(b) 80%
(c) 90%

(D) TRUE OR FALSE

1. The A climates receive their moisture from the ITCZ. (p.186)
T _____ F _____

2. The monsoon rainforest is often restricted to coastal areas backed by highlands. (p. 187)
T _____ F _____

3. The Maasai are a pastoral people living in the monsoon region of India. (p. 189)
T _____ F _____

4. A major contributing feature of the BS climate is its continentality. (p. 191)
T _____ F _____

5. Severe desertification has already affected an area the size of South America. (p. 193)
T _____ F _____

(E) SHORT-ANSWER QUESTIONS

1. What are the four sources of precipitation for the A climates? (p. 186)

____

____

2. What is the name given to the **Af climate** zone? (p. 186)

____

3. Why does the Af region have very little undergrowth? (p. 186)

____

4. What is the most significant feature of the annual march of temperature in the Af climate? (p. 186)

____

5. List five countries that have significant Af climate regions. (p.186)

____

____

6. Describe the nature of **shifting cultivation**. (p. 185)

____

____

7. During which season does the savanna experience most of its precipitation? (p. 180)

____

8. What type of vegetation is associated with the savanna region? (p. 181)

____

9. List five countries that are largely savanna. (p. 182)

____

10. What percent of humankind has been directly affected by **desertification**? (p. 193)

____

(F) MATCHING

Match the following terms and their meanings:

_______	1. climograph	a. daily variation
_______	2. deforestation	b. pastoral group in East Africa
_______	3. desertification	c. a graphical representation of temperature and precipitation
_______	4. diurnal	d. removal of forest cover
_______	5. fertigation	e. the southern edge of the Sahara
_______	6. Maasai	f. expansion of deserts
_______	7. Sahel	g. genetic engineering to allow plants to grow in brackish water

(G) ESSAY QUESTIONS

1. Name three areas on the west coasts of continents that have desert climate.

2. Explain why we have this west coast dry climate phenomenon.

3. Explain why the Patagonia region of Argentina experiences a dry climate.

4. Explain how changing perceptions of the environment have been largely responsible for rainforest depletion.

(H) GRAPHICACY

1. The unit opening photo on page 184 of the text is a dramatic illustration of very different climates side by side.

a. Why would this scene be a rarity in the realm of climate distribution?

b. What other setting would come closest to having such contrasts close together?

2. Figure 17-5 illustrates an Aw climate in the Southern Hemisphere. In the space allotted below, draw a climograph for an Aw climate in the Northern Hemisphere.

J F M A M J J A S O N D

a. What differences exist between the climograms for the different hemispheres?

(I) HYPOTHESIS CONSTRUCTION

1. List the forces that lead to rainforest depletion.

2. After rereading the sections on deforestation and desertification, write a response to the following statement.

"We should no longer speak of natural disasters; much of the tragedy of these events is in fact the result of social and economic forces."

3. Write a hypothesis about the causes of natural disasters that incorporates the ideas you have expressed above.

2. After reading the [illegible] on [illegible] write a response to the [illegible] this statement.

We should no longer speak of natural disasters [illegible] of the tragedy of these events is in fact the result of social and economic forces?

3. Write a hypothesis about the causes of natural disasters that incorporates the ideas you have expressed above.

Unit 18

Humid Mesothermal (C) Climates

(A) SUMMARY AND OBJECTIVES

The song that goes "It never rains in southern California" has a great tune, but it is right only half of the time. It does rain in southern California - but then only in the winter. This seasonal variation of rainfall is climate at its dramatic best. It allows southern California to develop a "dream factory" where movies can be shot outdoors for a large part of the year. Such a relationship between climate conditions and human activity is now a central feature of our discussion of the geography of climate. The C climates that are presented here are perhaps the most significant climates for humans because a large portion of the earth's population lives in these zones. These are the climates where humans harvest a large part of their food supply, but parts of this C climate zone also contains drought-prone areas that can surprise us with droughts such as happened in North American in 1988. In this unit we are again reminded that in areas of physical stress such as drought, social and economic factors do often play as important a role as the physical forces.

After studying this unit you should be able to:

1. Describe in detail the C climates.
2. Identify C climates from their respective climographs.
3. Distinguish the C subtypes.
4. Understand the impact of economic and social practices on the U.S. drought of 1988.

(B) KEY TERMS AND CONCEPTS

chaparral	Mediterranean climate
drought	mesothermal
Marine West Coast	topographic

(C) MULTIPLE CHOICE

_____ 1. In what portion of the continents is the humid subtropical climate situated? (p. 197)

(a) northwestern
(b) southern
(c) southeastern

_____ 2. What feature limits the eastward extent of Marine West Coast climates? (p. 199)
(a) rivers
(b) conservative politics
(c) mountain barriers

_____ 3. In what climate zone do we find the greatest seasonal variation of precipitation. (p. 199)
(a) Cfa
(b) Csa
(c) Cfb

_____ 4. What is the average annual precipitation range for Cs climates? (p. 200)
(a) 9 to 20 inches
(b) 14 to 54 inches
(c) 16 to 25 inches

_____ 5. What was the major cause of the U.S. drought of 1988? (p. 203)
(a) shifting jet stream
(b) reduction of cloud cover
(c) deforestation

(D) TRUE OR FALSE

1. In the C climates, temperature is a much more significant indicator than it is in the climates of the lower latitudes. (p. 195)
T _____ F _____

2. The C climate zone contains only a small portion of the world's population. (p. 196)
T _____ F _____

3. New Zealand is an example of a Marine West Coast climate. (p. 199)
T _____ F _____

4. Despite its reputation, London receives only 22 inches of precipitation per year. (p. 199)
T _____ F _____

5. Drought recurs in regular cycles. (p. 202)
T _____ F _____

(E) SHORT-ANSWER QUESTIONS

1. What are the temperature requirements for the mesothermal climates? (p. 195)

____

2. What locational feature ensures that the mesothermal climates have more changeable weather patterns than the tropical climates? (p. 195)

____

3. How do the Cs and the Cw climates differ? (p. 197)

____

4. What locational feature do the Cfa and the Cfb have in common? (p. 197)

____

5. What is the full name for the Cfa climate? (p. 197)

____

6. What are the temperature limits for the **Marine West Coast climate**? (p. 199)

____

7. What winds bring precipitation to the Cf climates? (p. 200)

____

8. What produces the markedly dry season in the **Cs climates**? (p. 200)

____

9. How does the Cw differ from the Aw climate? (p. 203)

____

10. Why does Cherrapunji, India have an exorbitantly high rainfall of 450 inches? (p. 203)

____

(F) MATCHING

Match the following places and their climates:

_______ 1. Cfa
_______ 2. Cfb
_______ 3. Csa
_______ 4. Cfc
_______ 5. Cwb

a. London, United Kingdom
b. Rekjavik
c. Cherrapunji, India
d. Charleston, South Carolina
e. Rome, Italy

(G) ESSAY QUESTIONS

1. Why does London, despite its low 22 inches of precipitation, still qualify as Cfb climate?

2. Describe how the U.S. drought of 1988 had as much of a social and economic cause as a physical one.

(H) GRAPHICACY

1. Refer to Figure 18-8.
   a. What is Reykjavik's latitude?

b. How does Reykjavik qualify as a Cfc climate?

2. Sketch a climograph for a Cs climate in the southern hemisphere.

J F M A M J J A S O N D

(I) HYPOTHESIS CONSTRUCTION

1. Figure 18-4 illustrates different landscapes in similar climatic regions.
   a. List the features which the two regions have in common.

   b. List the ways in which the differing landscapes are organized.

c. Name some reasons for the differences in landscape use and appearance.

d. Construct a hypothesis to explain such distinctively different imprints on landscape. Your independent variable should be "variations in landscape use".

Unit 19

Higher Latitude (D, E) and High-Altitude (H) Climates

(A) SUMMARY AND OBJECTIVES

How cold is it? It's so cold that the ice sticks to your tongue as you ask this question - that is if you happen to be in Verkhoyansk, Russia. The climates described in this unit are indeed chilly, but they can have an unexpected variation. In fact if you happen to be in the Arctic Circle in July, the sun never sets! Unit 19 describes the last of the Koppen climates, namely the colder, higher latitudes and highland regions. While these are the harshest and coldest climates, each contains subtypes that have developed unique adaptations between climate and vegetation so that limited agriculture is possible in many of them. While we may think of these climate zones as far away from most human beings, these previously pristine polar regions are now receiving much of the acid precipitation produced in the distant midlatitude zones. Ours is indeed a global interconnected system with activity in one area ultimately affecting regions far away.

After studying this unit you should be able to:

1. Understand the details of the higher latitude climates.
2. Interpret D and E climographs and locate these climates.
3. Explain the characteristics of highland climates.
4. Understand the nature of acid precipitation.

(B) KEY TERMS AND OBJECTIVES

acid precipitation
highland climate
icecap
microthermal
permafrost
taiga
tundra
vertical zonation

(C) MULTIPLE CHOICE

_____ 1. What percent of the earth's land area is covered by D climates? (p. 207)

(a) 11%
(b) 15%
(c) 21%

_____ 2. What does the word Taiga mean in Russian? (p. 207)

(a) Yeltsin
(b) snowforest
(c) lake

_____ 3. What is the name for the permanently frozen layer of sub-soil? (p. 208)

(a) thawsome
(b) permafrost
(c) perennial frost

_____ 4. What is the name given to the climate zone containing mosses, lichens, and stunted trees? (p. 210)

(a) taiga
(b) steppe
(c) tundra

_____ 5. On the pH scale what would a reading of 9 indicate?. (p. 212)

(a) acid
(b) alkaline
(c) neutral

(D) TRUE OR FALSE

1. The dry winter season of the Dw climate is a result of the inability of the cold air to hold enough moisture. (p. 207)
T _____ F _____

2. In the humid microthermal climates most of the precipitation comes in winter. (p. 209)
T _____ F _____

3. Almost all of the ET climates are found in the southern hemisphere. (p. 210)
T _____ F _____

4. Swarms of flies and mosquitoes are sometimes found in the Arctic zone. (p. 211)
T _____ F _____

5. At an altitude of 2,400 m the vegetation can resemble that of the Mediterranean zone. (p. 214)
T _____ F _____

(E) SHORT-ANSWER QUESTIONS

1. Explain the role of **continentality** in the D climates. (p. 207)

2. What are the temperature limits for the D climates? (p. 197)

3. What is the full name for the Dfa and Dwa climates? (p. 207)

4. In what way is the temperature at a place like Verkhoyansk distinctive? (p. 208)

5. How is the vegetation adapted to the low levels of insolation in the D climates? (p. 208)

6. Describe the major soil problem in the D climate area. (p. 208)

7. What is the distinctive seasonal feature of the **polar climates**? (p. 209)

8. What are the temperature characteristics of Tundra? (p. 210)

9. Name two places where the EF climate is found? (p. 211)

10. Define the concept **vertical zonation.** (p. 214)

____

____

(F) MATCHING

Match the following locations and their climate:

_______	1. Dwa	a. Moscow, Russia
_______	2. EF	b. Verkhoyansk, Russia
_______	3. Dfa	c. Churchill Falls, Labrador
_______	4. ET	d. Minneapolis, U.S.A.
_______	5. Dfc	e. Beijing, China
_______	6. Dfb	f. Point Barrow, Alaska, U.S.A.
_______	7. Dwd	g. Mirnyy, Antarctica

(G) ESSAY QUESTIONS

1. Discuss the problem of acid precipitation under the following headings.

a. The nature of acid precipitation.

b. The causes of acid precipitation.

c. The geographic extent of acid precipitation in the U.S.

(H) GRAPHICACY

1. Draw a sketch to indicate the nature of the pH scale.

2. Refer to Figure 19-1. What two distinctive features of this climograph tell us that this represents a Dw climate?

(I) HYPOTHESIS CONSTRUCTION

1. The climograph (Figure 19-5) for Barnaul, Russia has some similarity to the ET climograph (Figure 19-7).

a. Discuss the similarities and the differences.

b. Complete the sentence that will confirm the distinction between D and E climates.

The severity of the E climates is caused by ______________________________________

____

____

Unit 20

Dynamics of Climate Change

(A) SUMMARY AND OBJECTIVES

Even the tabloids can be right some of the time! Reports that the climate is changing have always been true but our real concern lies with the rate and direction of this change. We appear to be in a time of increasing global temperature and we have to consider the causes and implications of this planetary phenomenon. In this unit we are assured the agents creating this climate change are not aliens from another galaxy, but natural cycles in the changing relationship between the earth and the sun. Even without industrialization (which we consider in the next unit) it is clear that there have been cycles with higher and then lower temperatures. The scientific analysis of climate change is essential if we are to understand that our Earth thrives on change - it is surely not a boring planet.

After studying this unit you should be able to:

1. Understand the ever-present nature of climate change.
2. Evaluate the evidence of climate change.
3. Differentiate different climate periods in geological time.
4. Explain the different mechanisms for climate change.

(B) KEY TERMS AND CONCEPTS

climatic state
cryosphere
dendrochronology
interglaciation
isotopes
varves

(C) MULTIPLE CHOICE

_____ 1. What are samples of ocean sediment called? (p. 218)
(a) yields
(b) corn flakes
(c) cores

_____ 2. How many years ago did accurate recordkeeping of weather information begin? (p. 218)
(a) 400
(b) 150
(c) 200

_____ 3. How often do the sunspot cycles peak? (p. 223)

(a) every 26 years
(b) at every World Series
(c) every 11 years

_____ 4. What name is given to the cooling effect of the South Pacific Ocean? (p. 225)

(a) El Nino
(b) La Nina
(c) Zapata

_____ 5. What type of feedback is present when a snow-covered surface reduces atmospheric temperature? (p. 226)

(a) negative feedback
(b) neutral feedback
(c) positive feedback

(D) TRUE OR FALSE

1. Direct evidence of climatic change includes written historic records of crop yields. (p. 218)
T _____ F _____

2. Alternating layers of ocean sediment are called varves. (p. 218)
T _____ F _____

3. The 1940-1975 cooling trend was most strongly felt in Australia. (p. 220)
T _____ F _____

4. The Little Ice Age occurred about 750 years ago. (p. 221)
T _____ F _____

5. The earth's orbit around the sun changes in a cycle of 100,000 years. (p. 223)
T _____ F _____

(E) SHORT-ANSWER QUESTIONS

1. Describe what is meant by a **climatic state**. (p. 218)

____

____

2. What is meant by proxy climatic data? (p. 218)

____

3. What source provides the most detailed rock record of past climates? (p. 219)

____

4. When did the most recent glaciation of North America end? (p. 221)

____

5. What is the present glacial epoch known as? (p. 221)

____

6. How long did that previous interglaciation last? (p. 221)

____

7. Define the Late Cenozoic Ice Age. (p. 222)

____

8. What impact does the "wobble" of the Earth's axis have on climate? (p. 224)

____

9. Describe an example of a positive feedback on Earth's climate change. (p. 224)

____

10. What are "teleconnections" in the study of climate? (p. 226)

____

(F) MATCHING

Match the following terms and their meanings:

_______	1. cryosphere	a.	maximum period of climate warmth
_______	2. isotopes	b.	stormy areas on the surface of the sun
_______	3. dendrochronology	c.	lake bed sediments
_______	4. cores	d.	elements differentiated by the number of neutrons
_______	5. varves	e.	study of ice masses
_______	6. climatic optimum	f.	samples of earth sediment
_______	7. sunspots	g.	study of annual rings of trees to determine past events

(G) ESSAY QUESTIONS

1. What is the difference between historical climatic data and proxy climatic data?

2. Why are sediments left by glaciers of importance to earth scientists.

(H) GRAPHICACY

1. Refer to Figure 20-2. What conclusions can you draw about the weather during the period shown?

2. Refer to Figure 20-3. What tendency is indicated in this chart?

3. Refer to Figure 20-4.
   a. How many interglaciations have occurred in the last 1,000,000 years?

   b. What was the interval between these interglaciations?

(I) HYPOTHESIS CONSTRUCTION

1. At what angle from the perpendicular is the earth's axis tilted?

2. Speculate what climatic changes would occur if the earth's axis was tilted at an angle of 35^0.

3. What if the axis was tilted at an angle of only 10^0?

4. Complete the sentence.

The role of the particular earth axis angle is to ____

____

Unit 21

Human-Climate Interactions and Impacts

(A) SUMMARY AND OBJECTIVES

While aliens may not have changed our climates, urban growth and industrialization surely have. In this unit we come face-to-face with the paradox of our age - this is one of those Otoh Botoh times of life (on the one hand, but on the other hand). Cities have given us a great deal of cultural, economic, and social innovation and improvement. But they are also the areas where we are seeing the greatest negative impact that humans are having on the natural world. Increasing pollutants and changing surfaces have created urban microclimates with accompanying harsher weather and climate - more snow and rain exactly where we do not want it. The warning here is for us to monitor our impact and to make legislative and lifestyle changes that reduce these negative outputs. Remember those dinosaurs!

After studying this unit you should be able to:

1. Understand the concept of microclimates.
2. Identify the impacts that humans have on climate.
3. Explain the impact of urbanization on climate.
4. Evaluate the various types of air pollution.

(B) KEY TERMS AND CONCEPTS

dust dome	primary pollutants
microclimates	secondary pollutants
nuclear winter	urban heat island

(C) MULTIPLE CHOICE

_____ 1. What percent of the world's population lives in metropolitan areas? (p. 230)

(a) 50%
(b) 70%
(c) 90%

_____ 2. What factor has lead to the expansion of the Washington, D.C. urban heat island? (p. 231)

(a) industrialization
(b) congressional hot air
(c) rapid growth of population

_____ 3. By how much do urban areas have more precipitation than rural areas? (p. 233)

(a) 10%
(b) 20%
(c) 40%

_____ 4. What type of pollutant is referred to when sulfur dioxide changes to sulfur trioxide in the atmosphere? (p. 234)

(a) primary
(b) secondary
(c) tertiary

_____ 5. When did the dinosaurs become extinct? (p. 236)

(a) 100 million years ago
(b) 80 million years ago
(c) 65 million years ago

(D) TRUE OR FALSE

1. In the urban dust dome short wave radiation enters but long wave radiation is trapped inside. (p. 231)
T_____ F_____

2. Heat islands develop best under conditions associated with cyclonic systems. (p. 231)
T _____ F _____

3. The relative humidity is usually lower in cities than in rural areas. (p. 232)
T _____ F _____

4. Wind generally tends to have a higher velocity in urban areas than in rural areas. (p. 232)
T _____ F _____

5. La Porte, Indiana has had a dramatic change in its weather because it is located downwind from industrial Chicago. (p. 236)
T _____ F _____

(E) SHORT-ANSWER QUESTIONS

1. How does the heat capacity of urban materials influence temperature? (p. 231)

____

2. Describe an **urban dust dome**. (p. 231)

3. What are the best conditions for an **urban heat island** to develop? (p. 231)

4. Define **heat-island intensity**. (p. 232)

5. List the factors that determine the level of heat-island intensity. (p. 232)

6. What is the difference in rainfall occurrence between urban and rural areas? (p. 233)

7. Why are some pollutants called "status symbol" pollutants? (p. 234)

8. How is ground level-ozone created? (p. 234)

9. Why is London's air pollution in the 1950s often quoted as a classical case of severe weather modification? (p. 234)

10. Explain the nature of the **nuclear winter hypothesis**. (p. 235)

(F) MATCHING

Match the following terms and their meanings:

_______	1. isotherms	a. climate associated with a small area
_______	2. metabolic heat	b. dome-shaped layer of polluted air
_______	3. microclimates	c. metropolitan areas that have grown into each other
_______	4. dust dome	d. lines joining places of equal temperature
_______	5. conurbation	e. heat produced by the human body

(G) ESSAY QUESTIONS

1. Describe the energy flows of a city in systems terms.

2. Explain how urban areas differ from rural areas in their climate.

3. How can *relative* geographical location worsen the problem of urban air pollution?

4. What is the difference between the reducing and the oxidation types of secondary air pollution.

(H) GRAPHICACY

1. Refer to Figure 21-5. Describe the impact of the rivers on this urban heat island.

2. Draw an annotated cross-section of an urban heat island.

(I) HYPOTHESIS CONSTRUCTION

1. Table 21-1 notes the differences between urban and rural climates.

   a. Restate the hypothesis put forward by Stanley Changnon in 1968.

   b. Recount the evidence that supports this hypothesis.

PART THREE

THE BIOSPHERE

Unit 22

Climate, Soils, Plants, and Animals

(A) SUMMARY AND OBJECTIVES

Interaction the key word in this unit. Our concern is with the interaction between plants and soil, between climate and vegetation, between animals and plants. In fact this type of interaction is exactly what geography is all about. In Unit 1 we spoke about it in a theoretical sense, but here we deal with real world examples. The complexity of these interactions can be thought of as resembling a spider's web. Our task as scientists (detectives?) is to isolate certain linkages and analyze them in detail. One example might be: We have 30 cows in one pasture, how is the vegetation affected? At a more regional level, we ask questions about the U.S. Dustbowl and the depletion of species taking place right now. These outcomes are not inevitable and can be avoided if effective conservation efforts are in place. In this context, the geographer's approach is particularly valuable because it incorporates issues of scientific study with a concern for policy which will improve the care of our planet.

After studying this unit you should be able to:

1. Understand the concepts of natural geography and its subfields.
2. Appreciate the role of soils in the study of physical geography.
3. Explain the interaction between humans and the environment.
4. Discuss the nature of conservation and its significance for geography.

(B) KEY TERMS AND OBJECTIVES

biodiversity
biogeography
conservation
interface
phytogeography
species
The Dust Bowl
Wallace's Line

(C) MULTIPLE CHOICE

_____ 1. What term refers to geography of plants? (p. 241)
(a) pedology
(b) phytogeography
(c) zoogeography

_____ 2. What is the best indicator of the succession of prevailing ecosystems? (p. 242)

(a) climate
(b) changing dorm rooms
(c) vegetation

_____ 3. Of the world's 30 million species, how many have been identified and classified? (p. 242)

(a) 1.5 million
(b) 8.5 million
(c) 19.5 million

_____ 4. What do call animals whose young are born early and then carried in an abdominal pouch? (p. 243)

(a) mammals
(b) creotes
(c) marsupials

_____ 5. What is the name of the large regional project begun in 1933 to create electrical power in the southern U.S.? (p. 246)

(a) Georgia Power Company
(b) Tennessee Valley Authority
(c) Kentucky Utility Corps

(D) TRUE OR FALSE

1. The geography of soils is part of the science of pedology. (p. 241)
T _____ F _____

2. Soils may take about 1 to 2 years to develop enough to support permanent vegetation. (p. 241)
T _____ F _____

3. Alexander von Humboldt was the founder of biogeography as a distinct systematic field. (p. 243)
T _____ F _____

4. Alfred Wallace discovered that Australian animals existed not only in Australia but also in Africa. (p. 243)
T _____ F _____

5. Scientists have estimated that about one-quarter of all presently living plant species may become extinct during our lifetime. (p. 244)
T _____ F _____

(E) SHORT-ANSWER QUESTIONS

1. How does **natural geography** differ from the physical geography we have been studying up until now? (p. 243)

2. Why is the concept of **interface** so important for physical geographers? (p. 241)

3. Why does the moon's surface not contain soil? (p. 241)

4. What are the two goals of the geographical study of soils? (p. 242)

5. How have plant and animal domestication affected soil development? (p. 242)

6. Why did the discussion of Wallace's Line evoke such a great interest in the scientific community? (p. 244)

7. Define **conservation** as understood by the physical geographer. (p. 246)

8. Describe the environmental tragedy that occurred in the U.S. in the 1930s. (p. 246)

9. Name three states in which the Tennessee Valley Authority still operates. (p. 246)

10. What is the function of the Environmental Protection Agency? (p. 247)

____

____

(F) MATCHING

Match the following terms and their meanings:

_______	1. phytogeography	a.	study of distribution of plants and animals
_______	2. flora	b.	animal life
_______	3. pedology	c.	animals bearing young in pouches
_______	4. biogeography	d.	plant life
_______	5. fauna	e.	the geography of plants
_______	6. marsupials	f.	the study of soils

(G) ESSAY QUESTIONS

1. What role does water play in the soil forming process?

2. Why has Wallace's Line been so hotly debated?

3. Describe the phenomenon of the **Dust Bowl** in U.S. history.

4. Why is the concept of *natural geography* used less in geographical study today?

(H) GRAPHICACY

1. Refer to the unit opening photo on page 240 of the text. Identify three specific research questions evoked by the photograph which a geographer could pursue.

a.

b.

c.

(I) HYPOTHESIS CONSTRUCTION

1. What did Alfred Wallace discover in the course of his fieldwork in Australia and Southeast Asia?

2. What probable explanation can you give for this occurrence?

3. Construct a hypothesis to explain this phenomenon.

Unit 23

Formation of Soils

(A) SUMMARY AND OBJECTIVES

While it may be just dirt to you, soil is vital for human existence and it is arguably the most important part of our biosphere. After a description of the various organic and inorganic components of soil, we will shift attention to the factors that determine the soil-forming process. Note that the processes at work are rarely separate elements and in their interaction, they do provide us with a wide variety of soil types, all distinguished by particular soil profiles. After this unit and the next two, you may never look at soils in quite the same way again.

After studying this unit you should be able to:

1. Explain the various components of soil.
2. Understand the factors influencing soil formation.
3. List the processes at work in soil.
4. Describe and draw soil profiles.

(B) KEY TERMS AND CONCEPTS

elluviation	regolith
horizonation	soil profile
illuviation	soil
leaching	topography
minerals	weathering

(C) MULTIPLE CHOICE

_____ 1. What do we call the soil that forms directly from underlying rock? (p. 251)
(a) original soil
(b) parent soil
(c) residual soil

_____ 2. What are deposits of windblown material called? (p. 252)
(a) alluvium
(b) loess
(c) parmesan

_____ 3. Where does soil develop fastest? (p. 253)

(a) Indonesia
(b) Sweden
(c) Italy

_____ 4. Which soil horizon often receives dissolved and suspended particles from above? (p. 254)

(a) A
(b) B
(c) C

_____ 5. What are tiny mineral fragments (less than 0.1 microns in diameter) called? (p. 255)

(a) ions
(b) colloids
(c) anions

(D) TRUE OR FALSE

1. When rocks break down into soils, their mineral components become available as nutrients. (p. 251)
T _____ F _____

2. The soils in Georgia and in Maryland are similar despite their different climates. (p.252)
T _____ F _____

3. As biological soil agents, earthworms are much more efficient than ants and termites. (p. 252)
T _____ F _____

4. An organic horizon colored dark from vegetation growing in it is called an A horizon. (p. 254)
T _____ F _____

5. The C horizon is the soil layer in which the parent material is transformed by weathering. (p. 255)
T _____ F ______

(E) SHORT-ANSWER QUESTIONS

1. Why is soil regarded as a living system? (p. 249)

____

2. Why could there be no soil without an atmosphere? (p. 250)

____

3. Name the four components of soil. (p. 250)

____

4. Name two examples of organic matter. (p. 251)

____

5. What is the main function of water in the soil? (p. 251)

____

6. What is the origin of the red soils on the U.S. East Coast? (p. 251)

____

7. How does wind affect soil formation? (p. 252)

____

8. What is the role of bacteria in the soil? (p. 252)

____

9. Explain the idea of **soil regimes**. (p. 256)

____

____

10. How does climate influence aridisol soil formation? (p. 256)

____

____

(F) MATCHING

Match the following terms and their meanings:

_______	1. residual soil	a. windblown soil
_______	2. humus	b. differentiation of soil into distinct layers
_______	3. weathering	c. soil formed directly from underlying rock
_______	4. soil auger	d. partially decomposed organic matter
_______	5. loess	e. removal of valuable soil nutrients to lower levels
_______	6. soil horizon	f. disintegration of earth materials
_______	7. leaching	g. instrument used for boring holes in ground

(G) ESSAY QUESTIONS

1. What are the differences between renewable and nonrenewable resources.

2. Why could we classify soil as a *renewable* resource?

3. Compare the characteristics of residual and transported soils.

4. Describe the following soil processes noted by Simonson.

   a. Translocation

   b. Transformation

(H) GRAPHICACY

1. Draw an annotated sketch of an ideal soil profile, indicating horizons and their characteristics.

(I) HYPOTHESIS CONSTRUCTION

1. Compare and contrast the soil profiles presented in Figures 23-5 and 23-8 in the text.

2. Complete the following sentence.

The differences in soil characteristics noted in 1 above can be attributable to ________________

____

Unit 24

Physical Properties of Soil

(A) SUMMARY AND OBJECTIVES

If you've ever done any landscape work, you know for sure that soil is not the same in different locations. This knowledge of the geography of soil will be deepened by examining the physical and chemical characteristics of soil. After describing soil texture and structure, our discussion will shift to the nature and importance of soil acidity or alkalinity. Together with soil color and topography, these features play a prominent role in the regionalization and classification of soil that we will explore in Unit 25. Because soil is in a constant state of formation, the concluding section examines soil development in systems terms. Being able to correctly analyze and interpret the characteristics of soils can be beneficial to a farmer, a civil engineer, or a landscaper. We need that soil, so get acquainted.

After studying this unit you should be able to:

1. Understand the nature of soil texture and structure.
2. Interpret pH readings of soil.
3. Evaluate the impact of topography on soil development.
4. Explain in system terms how soil develops.

(B) KEY TERMS AND CONCEPTS

acidity	field capacity
alkaline	laterites
alluvium	loam
catena	silt
colluvium	solum

(C) MULTIPLE CHOICE

_____ 1. What soil characteristic is determined by the size of the soil particles? (p. 259)
(a) structure
(b) texture
(c) granulation

_____ 2. What do we call the ability of soil to hold water against the downward pull of gravity? (p. 260)

(a) hydrologic suction
(b) gravitational pull
(c) field capacity

_____ 3. What are soils called when their peds are arranged in columns? (p. 261)

(a) prismatic
(b) angular
(c) platy

_____ 4. What is often added to soils that are too high in alkalinity? (p. 263)

(a) sulfur
(b) lime
(c) pizza

_____ 5. What is a sequence of soil profiles appearing in regular succession? (p. 264)

(a) colluvium
(b) catena
(c) alluva

(D) TRUE OR FALSE

1. The soil solum consists of the A, B, and C horizons. (p. 259)
T _____ F _____

2. Silt has a greater field capacity than clay. (p. 260)
T _____ F _____

3. The greater the clay content, the more tightly the soil particles bind together. (p. 262)
T _____ F _____

4. In poorly drained soils, a thick B horizon develops. (p. 263)
T _____ F _____

5. Soils brought down by rivers are referred to as colluvium. (p. 264)
T _____ F _____

(E) SHORT-ANSWER QUESTIONS

1. Explain the term **solum.** (p. 259)

2. Which country has given us many of the world's leading soil scientists? Why? (p. 259)

3. What is the main activity in the C horizon? (p. 259)

4. How is a **pedon** different from a soil profile? (p. 259)

5. Rank the three major soil types by texture, from finest to coarsest. (p. 259)

6. What problems does clay soil pose for plants? (p. 260)

7. What is the water characteristic of granular structure? (p. 262)

8. What does the term **soil consistence** mean? (p. 262)

9. What oxides does a reddish soil probably contain? (p. 262)

10. Describe the nature of **colluvium**. (p. 263)

____

(F) MATCHING

Match the following terms and their meanings:

_______ 1. loam
_______ 2. solum
_______ 3. soil texture
_______ 4. pedon
_______ 5. soil structure
_______ 6. laterites
_______ 7. alluvium

a. soil deposited by streams
b. the A and B horizons containing active biological change
c. tropical soils having high iron oxide content
d. a soil containing equal amounts of silt, clay, and sand
e. the size of particles in the soil
f. a column of soil within a particular location
g. grouping of soil particles into clumps

(G) ESSAY QUESTIONS

1. What are the differences between the characteristics of platy and prismatic structure?

2. Imagine a weathered piece of rock material and a severed tree branch lying side by side. Describe briefly the processes that occur as these eventually form soil with attendant horizons.

3. Why is it important to determine the pH value of soil?

4. What is the approximate pH of the soil in your area?

(H) GRAPHICACY

1. Examine the unit opening photo on page 258.

   a. List the agricultural products most likely to thrive here.

   b. Why could this be a particularly good soil region?

   c. What part does culture play in the agricultural pursuits of the region in the photograph?

2. In the soil texture triangle (Figure 24-2), locate a soil that is
   a. 40% sand and 60% silt
   b. 70% clay and 30% sand

3. Which soil will be most porous? Explain.

(I) HYPOTHESIS CONSTRUCTION

1. Adding fertilizer to soil can have both positive and negative outcomes.

   a. Construct a hypothesis that could lead to finding support for the *positive* outcome of adding fertilizer.

   b. Construct a hypothesis that could lead to finding support for the *negative* outcome of adding fertilizer.

Unit 25

Classification and Mapping of Soils

(A) SUMMARY AND OBJECTIVES

The capstone of our study of soils is its classification and mapping. This unit reviews the history of recent attempts at soil classification and then it describes the details of the current widely-used Soil Taxonomy. The various Soil Orders are described with an emphasis on their chemical and physical properties as well as their potential for agriculture. Despite the large inputs of fertilizer used by agribusiness, we still see a strong correlation between areas of high agricultural productivity and particular soil types. Humans may work to increase crop yields but they cannot do it in a barren zone. In this sense then, our physical environment does pose some limits on our activities.

After studying this unit, you should be able to:

1. Understand the complexity involved in classifying soil.
2. Explain how the present Soil Taxonomy developed.
3. Distinguish the characteristics of the major soil orders.
4. Interpret soil classification maps.

(B) KEY TERMS AND CONCEPTS

alfisols
andisols
aridisols
entisols
histosols
inceptisols
mollisols
oxisols
soil order
spodosols
ultisols
vertisols

(C) MULTIPLE CHOICE

_____ 1. What are the major soils found in peat bogs? (p. 268)
(a) entisols
(b) alfisols
(c) histosols

_____ 2. What causes the swelling and cracking of the vertisols? (p. 269)
(a) water
(b) clay
(c) sand

_____ 3. Which soils contain large amounts of volcanic ash? (p. 271)
(a) histosols
(b) entisols
(c) andisols

_____ 4. Which soils cover the largest area of the earth's land surface? (p. 272)
(a) aridisols
(b) coffee grounds
(c) histosols

_____ 5. Which is the only soil order not found in the continental United States?. (p. 275)
(a) oxisols
(b) spodosols
(c) aridisols

(D) TRUE OR FALSE

1. Climate is not a strong influence on the distribution of entisols. (p. 268)
T _____ F _____

2. Inceptisols contain very little organic matter. (p. 269)
T _____ F _____

3. Mollisols are dominant in the U.S. Great Plains and in south-central Russia. (p. 272)
T _____ F _____

4. Alfisols are usually dry soils. (p. 273)
T _____ F _____

5. In areas of spodosols, lumbering is often a major economic activity. (p. 273)
T _____ F _____

(E) SHORT-ANSWER QUESTIONS

1. What was Vasily Dokuchayev's initial insight that changed the way soils were studied? (p. 266)

____

2. Why was the Marbut System not adequate for classifying soils? (p. 267)

____

3. What is the **Seventh Approximation**? (p. 267)

____

4. What crops can best be grown in histosols? (p. 269)

____

5. What impact can vertisols have on human activity? (p. 269)

____

6. How do alfisols differ from mollisols? (p. 272)

____

7. Name three crops grown in alfisols. (p. 273)

____

8. Describe the process which results in spodosols. (p. 273)

____

9. Where are oxisols found? (p. 274)

____

10. Describe the major human activity in oxisol regions. (p. 274)

____

(F) MATCHING

Match the following terms and their meanings:

_______ 1. alfisols
_______ 2. histosols
_______ 3. spodosols
_______ 4. aridisols
_______ 5. inceptisols
_______ 6. oxisols
_______ 7. mollisols

a. youthful soils with poorly developed horizons
b. soils having thin profile with little organic material
c. leached soils of the tropics
d. organic, water-saturated soils
e. soils rich in humus and clay
f. moist and high in mineral content soils
g. acidy soils with depleted A horizon material

(G) ESSAY QUESTIONS

1. Distinguish between the concepts of *Family* and *Series* in the Soil Taxonomy.

2. Explain why the hypothetical climate map (on page 178 of the textbook) has strong similarities to the hypothetical soil distribution patterns on page 268.

3. Why are mollisols so significant for agriculture?

(H) GRAPHICACY

1. Examine Figure 25-13 and answer the following questions:

   a. Describe and explain the location of the inceptisols.

   b. Describe and explain the location of spodosols.

2. Compare and contrast the soil profiles of mollisols (Figure 25-8) with oxisols (Figure 25-11).

(I) HYPOTHESIS CONSTRUCTION

1. Compare the map of soil classification (Figure 25-14) with the climate classification map (Figure 16-3).

a. Which soil system is most closely associated with a particular climate?

b. Construct a hypothesis to explain this relationship.

Unit 26

Biogeographic Processes

(A) SUMMARY AND OBJECTIVES

What is the connection between a whale, a seal, and a tiny fish? If you said that one feeds on the other which in turn feeds on the third, then you would be correct. It is this chain of connections that we will examine in this unit. The processes at work range from the microscopic activity of photosynthesis, to the cow eating the grass, and to humans having their summer barbecues. Uncovering these biological activities and then determining their connections, is one of the most exciting tasks of the biogeographer. It is in some ways a detective tale about who did what to whom, where? In the concluding section our attention will turn to the WHERE question - the geography of this biological world. Our task is to determine if there are patterns in the distribution of species and then to ask how these are being influenced by the natural environment and, increasingly, by human activity.

After studying this unit you should be able to:

1. Understand the role of photosynthesis in the biosphere.
2. Explain the natural environment in ecosystem terms.
3. Understand the concepts of food chain and ecological efficiency.
4. Describe the factors influencing the distribution of species.

(B) KEY TERMS AND CONCEPTS

biomass
chlorophyll
climax vegetation
ecosystem
food chain
herbivores
photosynthesis
phytogeography
phytomass
plant succession
trophic level
zoogeography

(C) MULTIPLE CHOICE

_____ 1. In what zone is photosynthesis most active? (p. 280)
(a) polar areas
(b) tropics
(c) midlatitude zones

_____ 2. What is the name for plant-eating animals? (p. 281)

(a) carnivores
(b) herbivores
(c) grassivores

_____ 3. Approximately what percent of the energy produced in the form of food is passed from one trophic level to another? (p. 282)

(a) 10%
(b) 20%
(c) 30%

_____ 4. What do we call the plant succession when the vegetation is in complete harmony with all elements of it environment? (p. 282)

(a) autogenic succession
(b) allogenic succession
(c) climax community

_____ 5. What do we call plants that are adapted to heat?. (p. 284)

(a) megatherms
(b) party animals
(c) microtherms

(D) TRUE OR FALSE

1. Photosynthesis removes oxygen from the air. (p. 279)
T _____ F _____

2. Tropical swamps have the highest efficiency of all ecosystems. (p. 282)
T _____ F _____

3. An example of linear autogenic succession is when one kind of vegetation is replaced by another, and that, in turn, is replaced by the original. (p. 282)
T _____ F _____

4. Plants adapted to dry areas are called xerophytes. (p. 284)
T _____ F _____

5. Mutualism refers to a biological interaction where one species inhibits the spread of another. (p. 285)
T _____ F _____

(E) SHORT-ANSWER QUESTIONS

1. What are the three elements required for photosynthesis? (p. 279)

2. What is the atmospheric function of photosynthesis? (p. 279)

3. Explain how **evapotranspiration** differs from evaporation. (p. 280)

4. Define **ecosystem**. (p. 280)

5. Describe a **trophic level** using one example to illustrate your answer. (p. 281)

6. What is meant by the **optimum range** for species. (p. 283)

7. Distinguish between hygrophytes and mesophytes. (p. 284)

8. Name four edaphic factors which influence plant dispersal. (p. 284)

9. Describe an example of term **mutualism** in plant survival. (p. 286)

____

____

10. Explain the concept **species-richness gradient**. (p. 287)

____

____

(F) MATCHING

Match the following terms and their meanings:

_______	1. biomass	a.	when vegetation is in harmony with the environment
_______	2. climax community	b.	plants that are adapted to heat
_______	3. chlorophyll	c.	microscopic green plants
_______	4. phytoplankton	d.	plants that are adapted to dry areas
_______	5. stomata	e.	total living plant and animal matter in an area
_______	6. xerophytes	f.	the green pigment in plants
_______	7. megatherms	g.	small holes in the leaf surface which allow moisture to pass through

(G) ESSAY QUESTIONS

1. Why has the concept of ecology become so important recently?.

2. What is the relationship between photosynthesis and the green color of most vegetation.

3. What is meant by *ecological efficiency*.

4. What is the difference between linear autogenic and allogenic succession.

(H) GRAPHICACY

1. Compare the map of biomass production (Figure 26-3) with the map of the Koppen climate system (Figure 16-3) and describe the relationship between these patterns as they appear in Africa.

2. Examine Figure 26-11 and explain this scene in terms of ecological efficiency.

(I) HYPOTHESIS CONSTRUCTION

1. Reread the Perspective box on the Species-Richness Gradient (page 287 in the textbook).

   a. Write down the main hypothesis contained in this piece.

   b. Write down a major secondary hypothesis which is discussed in the section.

Unit 27

The Global Distribution of Plants

(A) SUMMARY AND OBJECTIVES

Any visitor from outer space would definitely be astounded at the great variety of vegetation found in different parts of the world. The visitor would undoubtedly marvel at a planet that can simultaneously support a vast stretch of rainforest as well as a few scattered pine trees in a snow-bound ice desert. In this unit we examine the characteristics of various plant types within their particular spatial contexts. Again we note that a single physical feature (plants in this case) never stands quite alone, but is always part of a complex unity of relationships, including the climate, soils, topography, etc. The web of life ties it all together.

After studying this unit you should be able to:

1. Understand the biome concept.
2. Identify the characteristics of specific biomes.
3. Recognize the biological integration within biomes.
4. Explain the global distribution of biomes and their relationship to human activity.

(B) KEY TERMS AND CONCEPTS

biome	epiphytes
boreal	perennials
chaparral	scrub
deciduous	succulents

(C) MULTIPLE CHOICE

_____ 1. What are the most common trees of the savanna region? (p. 294)
(a) evergreen
(b) coniferous
(c) deciduous

_____ 2. What two soil types are most common in the desert biome? (p. 294)
(a) histosols and aridisols
(b) aridisols and entisols
(c) entisols and vertisols

_____ 3. What biome is present in the Pampas region of Argentina? (p. 295)
(a) temperate grassland
(b) tango parties
(c) savanna

_____ 4. In which Mediterranean biome does one find a chaparral landscape. (p. 296)
(a) South Africa
(b) Spain
(c) California

_____ 5. Where are the boreal forests found? (p. 296)
(a) southern part of Australia
(b) northern part of Canada
(c) southern part of France

(D) TRUE OR FALSE

1. In a tropical rainforest, only about 1% of the sunlight above the forest actually reaches the ground. (p. 292)
T ____ F ____

2. Temperate forests have only about one quarter of the number of species per acre than tropical rainforests do. (p. 292)
T _____ F _____

3. The temperate grassland occurs on the east coasts of continents. (p. 294)
T _____ F _____

4. Lianas are vines rooted in the ground with flowers in the forest canopy. (p. 292)
T _____ F _____

5. Monsoon rainforests have greater variety of plant species than tropical rainforests. (p. 293)
T _____ F _____

(E) SHORT-ANSWER QUESTIONS

1. Define a **biome**. (p. 289)

____

____

2. What are the two major factors determining biome distribution? (p. 290)

____

3. Where does undergrowth flourish in tropical rainforests? (p. 292)

____

4. What impact does **leaching** have on tropical rainforests? (p. 292)

____

5. What fauna dominates the food chain in the savanna biome? (p. 294)

____

6. Name two areas that are comprised of temperate grassland biome. (p. 295)

____

7. What are the two common soil orders found in temperate grassland regions? (p. 295)

____

8. What is ***fynbos***? (p. 296)

____

9. Name two tree types found in the northern coniferous forest biome. (p. 296)

____

10. What type of vegetation is found in the tundra? (p. 297)

____

(F) MATCHING

Match the following terms and their meanings:

_______	1. perennials	a.	plants that use trees for support
_______	2. lianas	b.	plants completing their life cycle in a single growing season
_______	3. succulents	c.	plants persisting from year to year
_______	4. chaparral	d.	snowforests in Russia
_______	5. annuals	e.	vines rooted in the ground
_______	6. epiphytes	f.	shrubby Mediterranean vegetation in California
_______	7. taiga	g.	plants with fleshy, water-storing leaves

(G) ESSAY QUESTIONS

1. Describe the key factors that determine the distribution of biomes.

2. How does relief influence biome distribution?

(H) GRAPHICACY

1. Draw in the tropical savanna and the temperate grassland biomes on the following map.

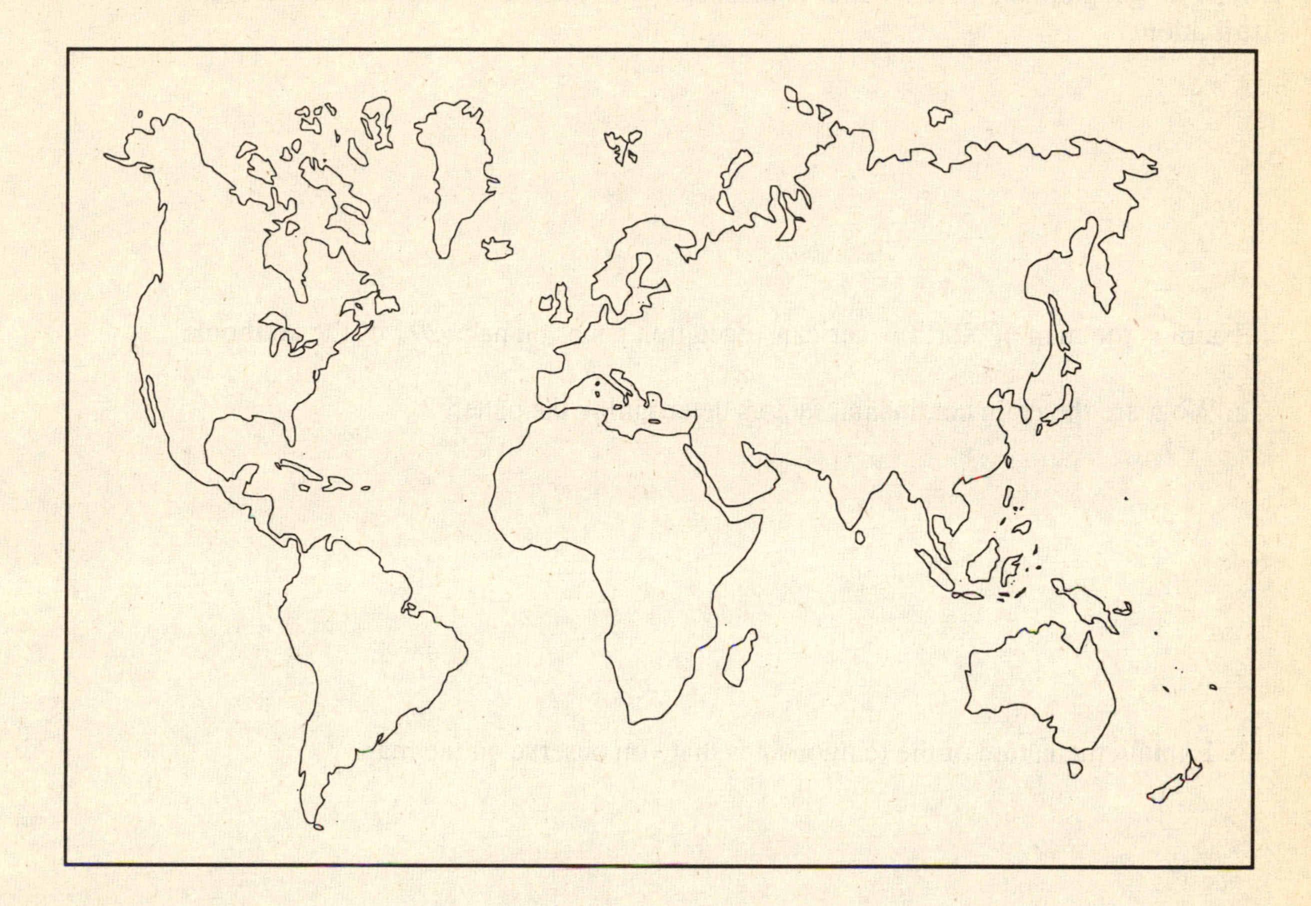

2. Describe the main locational features of the tropical savanna and the temperate grassland as sketched.

(I) HYPOTHESIS CONSTRUCTION

1. Why do geographers have a better understanding of plant distributions than animal distributions?

2. Examine the map of North American vegetation zones on page 291 of the textbook.

   a. What are the main continental factors determining the pattern?

   b. Explain the nature of the relationships that you observe on the map.

Unit 28

Zoogeography: Spatial Aspects of Animal Populations

(A) SUMMARY AND OBJECTIVES

Congratulations! You have just been named head of the World Wildlife Organization! You're helping create a better world (and probably earning a good salary as well). Now, what is your first task? Pssst, just a hint. Why not hire some biogeographers. That's right - you need to know WHERE the animals are before you can make any policy about them. The board will also be very impressed by those colorful maps showing animal distributions. Your commissioned geographical study of animal populations should include the basic ideas of evolution, the concepts of mutation, ecological niche, and adaptation. After the team has described the various zoogeographic realms, you should focus attention on the increasing human imprint on animal habitats. Only through incorporating this integrative zoogeographic approach, can your preservation efforts hope to effectively stop the depletion of the earth's animal and plant resources. Good luck in your new job.

After studying this unit you should be able to:

1. Understand the role of evolution in the animal world.
2. Explain the basic concepts of zoogeography.
3. Identify the various zoogeographic realms.
4. Evaluate the nature of human impact on animal habitats.

(B) KEY TERMS AND CONCEPTS

animal ranges	mutation
ecological niche	Nearctic
habitat	realms

(C) MULTIPLE CHOICE

_____ 1. What do we call the place where a species can best sustain itself and thrive? (p. 299)
(a) realm
(b) ecological niche
(c) a commune

_____ 2. The Serengeti Plain is an example of what type of habitat? (p. 300)
(a) desert
(b) tundra
(c) savanna

_____ 3. Who published the classical work on evolution *The Origin of Species*? (p. 300)
(a) Alex von Humboldt
(b) Jerry Miles
(c) Charles Darwin

_____ 4. Which realm contains the earth's most varied fauna? (p. 302)
(a) neotropic
(b) paleotropic
(c) palearctic

_____ 5. What is the difference between the Australian and the New Zealand realm? (p.302)
(a) Australia has no mammals
(b) Australians have better barbecues
(c) New Zealand has no mammals

(D) TRUE OR FALSE

1. Most niches are no larger than a few square feet. (p. 299)
T _____ F _____

2.) Habitats may contain many niches within them. (p. 299)
T _____ F _____

3.) It is easier to draw maps of faunal regions than of floral distributions. (p. 301)
T _____ F _____

4.) The nearctic and the palearctic realms are much less diverse in fauna than the other realms. (p. 302)
T _____ F _____

5. It is not the size of an island, but rather its internal variability that affects the species total. (p. 303)
T _____ F _____

(E) SHORT-ANSWER QUESTIONS

1. What is the function of a specialized niche? (p. 299)

2.) Distinguish between a niche and a **habitat**. (p. 299)

3. What is an **archipelago**? (p. 300)

4. When compared to plant areas, why is it so difficult to accurately indicate faunal regions? (p. 301)

5. What was the difference between the Wallace and the Weber Lines? (p. 300)

6. Name two specific physical features that may be a boundary between zoogeographic realms. (p. 301)

7. In what ways are New Zealand and Australia's fauna similar. (p. 302)

8. Describe the concept **convergent evolution**. (p. 302)

9. What determines the range of the spotted owl? (p. 305)

____

10. Name two reasons why the south Asian fauna have been severely decimated. (p. 306)

____

(F) MATCHING

Match the following terms and their meanings:

_______	1. habitats	a. an inheritable change in the DNA of a gene
_______	2. Neotropic realm	b. the environment occupied by a species
_______	3. Palearctic realm	c. basic physical unit of heredity, carrying information from parents
_______	4. mutation	d. zoogeographic region of South America
_______	5. Nearctic realm	e. zoogeographic region of Asia, Europe, and North Africa
_______	6. genes	f. zoogeographic region of North America and Greenland

(G) ESSAY QUESTIONS

1. What particular approach have geographers such as Darlington and Simpson taken when discussing animals and evolution?

2. Explain the nature of adaptation as an evolutionary principle.

3. Describe the Serengeti Plain as an ecosystem.

(H) GRAPHICACY

1. World map exercise.

   a. Locate and circle the Serengeti Plain, the Galapagos Islands, and Madagascar on the map.

   b. Draw in both the Wallace and the Weber Line.

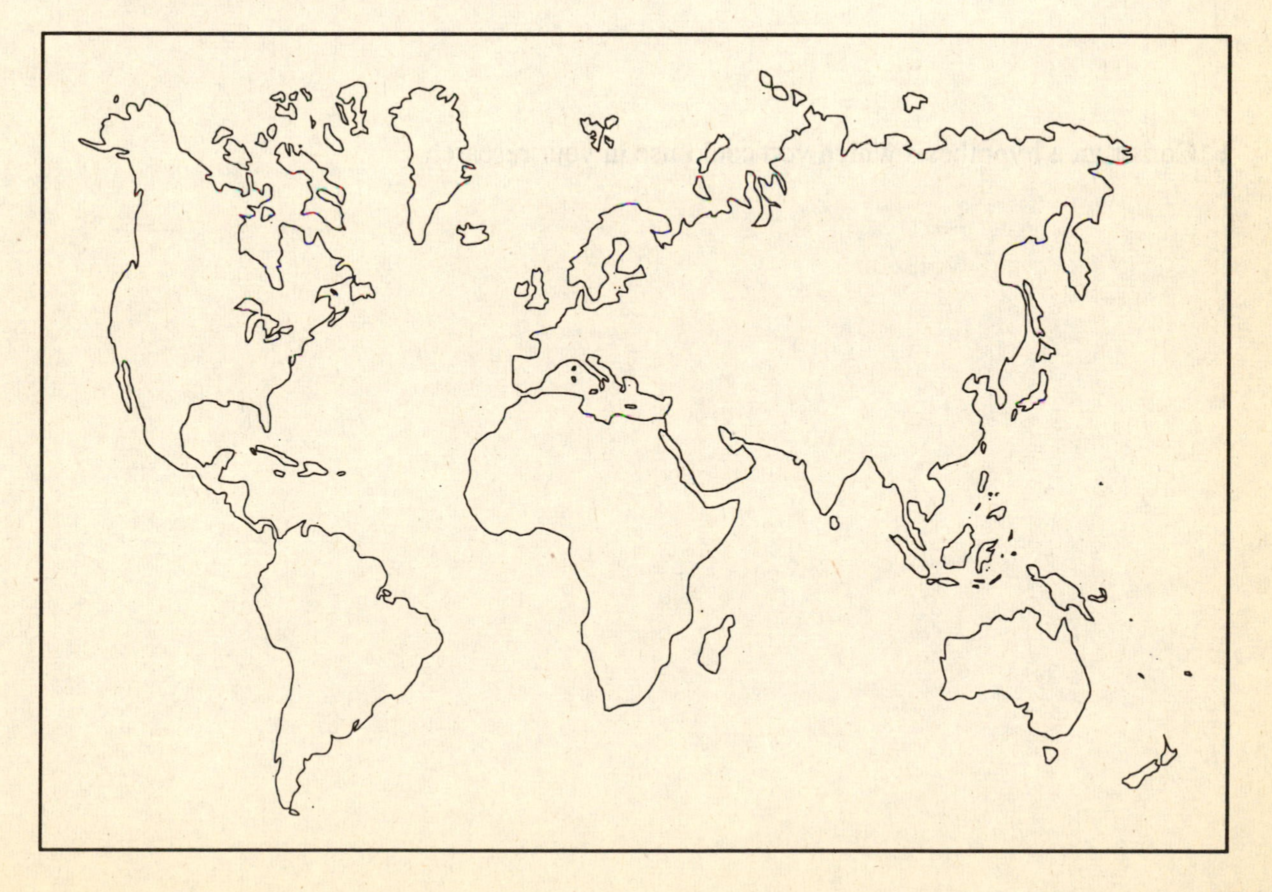

c. In which other areas of the world could the Wallace/Weber Line approach be applicable?

(I) HYPOTHESIS CONSTRUCTION

1. Madagascar has a very different zoogeographic mix than the nearby African continent. You have just been given a grant to study this phenomenon.

a. Suggest some reasons for Madagascar's unique mixture of animals..

b. Construct a hypothesis which you could use in your research.

PART FOUR

THE RESTLESS CRUST

Unit 29

Planet Earth in Profile: The Layered Interior

(A) SUMMARY AND OBJECTIVES

Journey to the Center of the Earth, a great science fiction story by Jules Verne, tells about adventurers trying to reach the earth's core. While this may be a fantasy tale (they never did reach the earth's center), the story illustrates many questions which scientists have long had about the interior of the earth. In this unit we explore the earth's internal structure and the origin of seismic tremors that can cause so much damage on our surface. While much of the interior of the earth is not directly accessible, we spend time analyzing the earth's crust because even here much scientific work needs to be done. Many of the present researchers were first attracted to scientific discovery by the writings of Jules Verne. It may have been science fiction, but in 1865 Verne predicted a landing on the moon - 104 years before it actually happened!

After studying this unit you should be able to:

1. Describe the interior structure of the earth.
2. Explain the characteristics of the earth's crust, lithosphere, and mantle.
3. Locate the major world continental relief systems.
4. Understand the nature of gradational processes.

(B) KEY TERMS AND CONCEPTS

asthenosphere	orogeny
lithosphere	seismic waves
mantle	sial
Moho discontinuity	sima

(C) MULTIPLE CHOICE

_____1. How far into the earth have the deepest boreholes penetrated? (p. 310)
(a) 3 km
(b) 8 km
(c) 12 km

____ 2. Which seismic waves move material in their path parallel to the direction of movement? (p. 312)

(a) S
(b) P
(c) L

____ 3. What are the two main elements found in the earth's core? (p. 313)

(a) iron and magnesium
(b) magnesium and carbon
(c) iron and nickel

____ 4. What is the range of the thickness of the crust? (p. 315)

(a) 5 to 40 km
(b) 2 to 80 km
(c) 15 to 70 km

_____ 5. What is the name of the continental shield that covers the northern part of North America? (p. 318)

(a) Stephanie shield
(b) Siberian shield
(c) Laurentide shield

(D) TRUE OR FALSE

1. While elements such as iron and magnesium are rarely found in the earth's crust, they are concentrated in the deeper layers. (p. 311)
T _____ F ____

2. The speed of seismic waves stays the same from its origin to the surface. (p. 311)
T _____ F _____

3. The rocks that make up the continental landmasses have a lower density than the oceanic crust. (p. 315)
T _____ F _____

4. The crust is a continuous layer within the earth's structure. (p. 316)
T _____ F _____

5. The soft plastic layer of the upper mantle is called the asthenosphere. (p. 316)
T ____ F ____

(E) SHORT-ANSWER QUESTIONS

1. What are the two main objectives when deep-earth samples are studied? (p. 311)

2. Where in the earth are the heavy elements located? (p. 311)

3. Distinguish between **seismic reflection** and **seismic refraction.** (p. 311)

4. Describe the contact zone between the core and the mantle. (p. 313)

5. Define the **Moho discontinuity**. (p. 314)

6. Describe the varying thickness of the crust. (p. 315)

7. Distinguish between **sial** and **sima** in the crust. (p. 315)

8. Explain the nature of **lithospheric plates.** (p. 316)

9. Define **gradational processes.** (p. 319)

10. Distinguish between **weathering** and **erosion**. (p. 320)

____

____

(F) MATCHING

Match the following terms and their meanings:

_______ 1. seismic waves
_______ 2. lithosphere
_______ 3. seismograph
_______ 4. sialic rocks
_______ 5. relief
_______ 6. asthenosphere
_______ 7. orogenic belts

a. granitic continental rocks of silica and aluminum
b. a series of linear mountain chains
c. vertical difference between the highest and lowest elevation
d. soft, plastic layer in the upper mantle
e. instrument for measuring earthquakes
f. solid upper part of the mantle and the crust
g. pulses of energy generated by earthquakes

(G) ESSAY QUESTIONS

1. Describe the differences between *L*, *P*, and *S* seismic waves.

2. Explain why knowledge of the Moho discontinuity is so important to earth scientists.

(H) GRAPHICACY

1. On the map of North America, draw in the following physiographic features:
Laurentide Shield; Rocky Mountains; Appalachians; Central Plains; the Mississippi Valley.

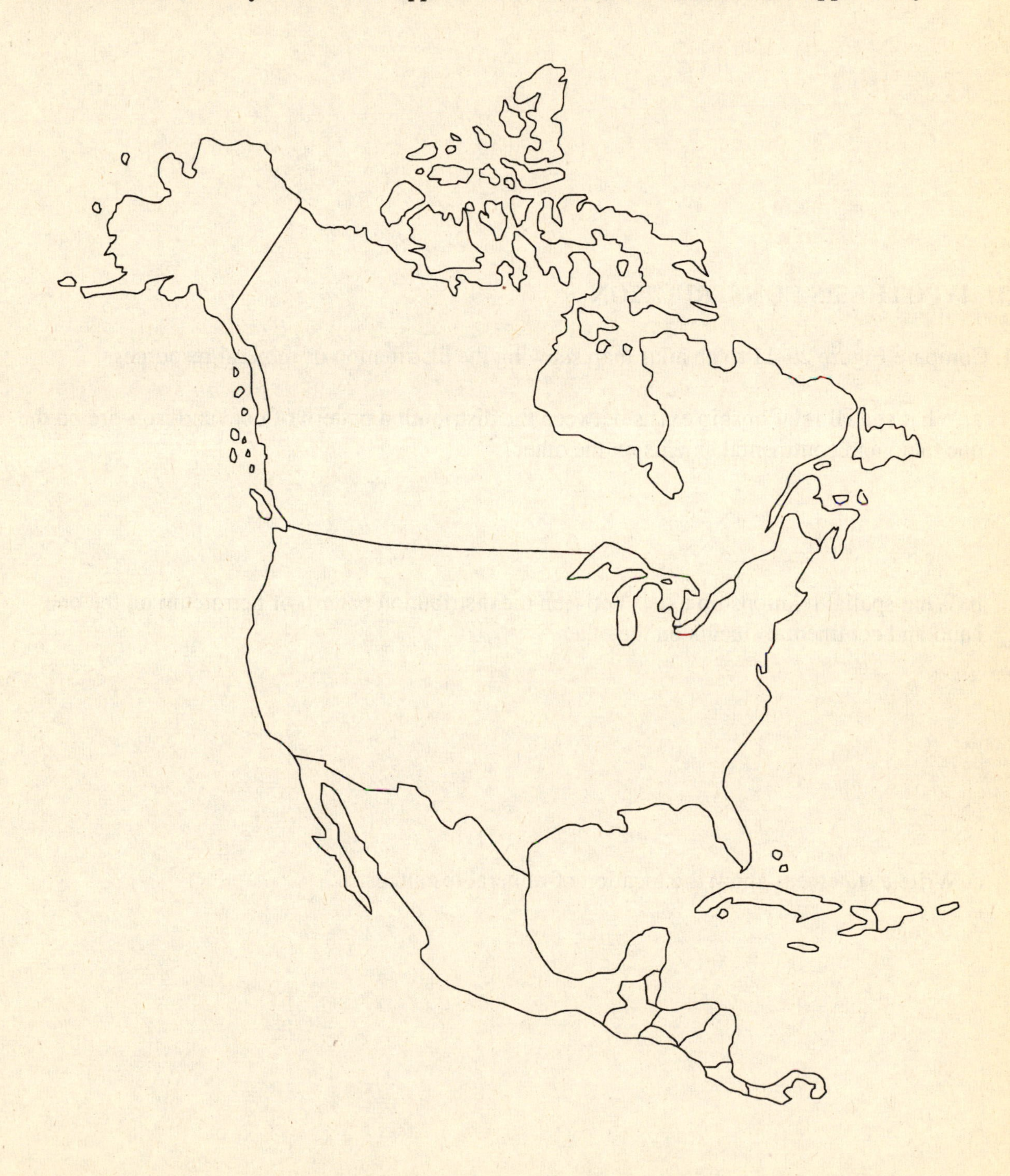

a. Describe how the physiography displayed on this map has influenced human settlement and activity.

(I) HYPOTHESIS CONSTRUCTION

1. Compare Figure 29-11 to an atlas map showing the distribution of mineral resources.

   a. What spatial relationship exists between the distribution pattern of coal and iron-ore on the one hand and continental shields on the other?

   b. What spatial relationship exists between the distribution pattern of petroleum on the one hand and continental shields on the other?

   c. Write a statement about the location of mineral resources.

Unit 30

Minerals and Igneous Rocks

(A) SUMMARY AND OBJECTIVES

Rocks are not just rocks! They are as varied as the palms of our hands and as interesting. On our way to understanding the surface processes that create our soil, our agriculture, and our corn-on-the-cob, we must first know about the structure of rock types. Rocks are assemblages of minerals and are form the critical factor in our discussion of the solid earth. The first rock type described is the igneous variety which originates deep inside the earth and often finds its way to the surface in the form of volcanoes, plateaus, and domes. Geographers have a particular interest in this topic because rocks constitute not only the background to our landscapes, but also because these physical landscapes can and do shape human activity in particular ways.

After studying this unit you should be able to:

1. Understand the relationship between rocks and minerals.
2. Explain the characteristics of minerals.
3. Distinguish between igneous, sedimentary, and metamorphic rocks.
4. Describe the occurrence of igneous rocks in the landscape.

(B) KEY TERMS AND CONCEPTS

batholith	lava
extrusive	magma
igneous rocks	minerals
jointing	rocks

(C) MULTIPLE CHOICE

_____ 1. On the hardness scale, which mineral is the hardest? (p. 323)
(a) talc
(b) calcite
(c) quartz

_____ 2. What is formed when carbonates combine with calcium? (p. 324)
(a) calcite
(b) quartz
(c) aluminum

_____ 3. What we do we call a vertical intrusive layer of magma? (p. 325)
(a) batholith
(b) dike
(c) sill

_____ 4. Which rock type exhibits a rectangular jointing pattern? (p. 326)
(a) granite
(b) beryl
(c) feldspar

_____ 5. What type of physical feature is represented by Ship Rock in New Mexico? (p. 327)
(a) exfoliation
(b) Clint Eastwood's profile
(c) dike

(D) TRUE OR FALSE

1. Minerals can be distinguished by the shape of their crystals and by their color and hardness. (p. 322)
T ____ F _____

2. About 54 minerals make up more than 98% of the earth's crust by weight. (p. 324)
T ____ F _____

3. Intrusive igneous rocks tend to have iarger mineral crystals than extrusive rocks. (p. 325)
T ____ F _____

4. Original magma which is rich in silica yields feldspar and quartz. (p. 325)
T ____ F ____

5. Jointing is only found in igneous rocks. (p. 327)
T ____ F ____

(E) SHORT-ANSWER QUESTIONS

1. What is the difference between elements and minerals? (p. 321)

____

____

2. Distinguish between rock **streak** and **luster**. (p. 323)

____

____

3. Why are carbonates of such great interest in physical geography? (p. 324)

____

4. Define **igneous rocks**. (p. 324)

____

5. Distinguish between intrusive and extrusive igneous rocks. (p. 325)

____

____

6. Name two fine-grained extrusive igneous rocks. (p. 325)

____

7. Distinguish between **concordant** and **disconcordant** intrusions. (p. 325)

____

____

8. Describe **sills** and **dikes**. (p. 325)

____

____

9. How does a **laccolith** differ from a batholith? (p. 325)

____

____

10. Define **jointing.** (p. 326)

____

____

(F) MATCHING

Match the following terms and their meanings:

_______	1. magma	a. the tendency of minerals to break in certain directions
_______	2. plutonic rocks	b. magma that reaches the earth's surface
_______	3. obsidian	c. a table-like landform
_______	4. cleavage	d. coarse-grained intrusive rocks
_______	5. lava	e. a massive body of plutonic rock
_______	6. mesa	f. molten rock
_______	7. batholith	g. black extrusive rock

(G) ESSAY QUESTIONS

1. Describe the Mohs' Hardness Scale and name examples on opposite ends of the scale.

2. What are the differences between sedimentary and metamorphic rocks?

3. How can we determine the origin of a particular igneous rock by noting its color?

(H) GRAPHICACY

1. Landforms have always featured prominently in movies, either as part of the story or serving as the backdrop to a plot.

a. In which movie did Wyoming's Devil's Tower (photo on page 327 in the textbook) play a major role.

b. Can you name any other movies which feature specific physical landforms. Describe those landforms.

(I) HYPOTHESIS CONSTRUCTION

1. From which direction is the unit opening photo of Mt. St. Helens on page 321 in the text taken? Explain your answer.

2. What are the common features connecting the landscape of Rio de Janeiro in Figure 30-6 and Mt. St. Helens in the opening photo on page 321?

3. Describe how the Rio landscape may have developed.

Unit 31

Sedimentary and Metamorphic Rocks

(A) SUMMARY AND OBJECTIVES

In this unit we examine sedimentary and metamorphic rocks and their relationships in the rock cycle. Sedimentary rocks are especially common in their layered appearance and in their ability to contain petroleum. In the study of metamorphic rocks, we are reminded that change is a significant feature of earth's inorganic material. The possibility of heat and pressure causing rock to change was demonstrated the character Superman in his attempt to change coal into diamond. This may be taking dramatic license, but the general idea is fine. Nothing stays quite the same in our constantly changing planet. The photographs in the textbook illustrate this process-oriented approach and help identify the rock types.

After studying this unit you should be able to:

1. Explain the characteristics of sedimentation.
2. Understand the processes leading to metamorphic rocks.
3. Identify rocks as sedimentary or metamorphic.
4. Understand the premise of the rock cycle.

(B) KEY TERMS AND CONCEPTS

clastic	metamorphism
compaction	rock cycle
conglomerate	sandstone
lithification	stratification

(C) MULTIPLE CHOICE

_____1. What name is given to pebble-sized rock fragments that are angular and jagged? (p. 330)
(a) conglomerate
(b) clastics
(c) breccia

_____ 2. Which of these are sedimentary rocks? (p. 331)
(a) shale
(b) slate
(c) gneiss

____ 3. What do we call the contact between eroded strata and the strata of resumed deposition? (p. 332)

(a) stratigraphy
(b) bedding
(c) unconformity

____ 4. What is the name of the process that gives metamorphic rocks their banded appearance? (p. 334)

(a) foliation
(b) schisting
(c) dessication

______ 5. What is the parent material from which marble developed? (p. 334)

(a) schist
(b) limestone
(c) sandstone

(D) TRUE OR FALSE

1.The vast majority of sedimentary rocks are clastic in nature. (p. 330)
T ____ F ____

2. Most limestone results from the distilling of calcium carbonate from seawater. (p. 332)
T ____ F _____

3. Sedimentary rocks contain very few fossils. (p. 332)
T ____ F _____

4. Quartzite is a very hard rock that resists weathering.
(p. 334)
T ____ F _____

5. Unlike shale, slate does not break along parallel lines.
(p. 334)
T _____ F _____

(E) SHORT-ANSWER QUESTIONS

1. Why are sedimentary and metamorphic rocks referred to as **secondary rocks**? (p. 329)

___

2. Describe the process of **compaction**. (p. 329)

___

3. How do **nonclastic** sediments differ from clastic ones? (p. 330)

___

4. What is characteristic about the shape of conglomerate components? (p. 330)

___

5. How does the composition of shale differ from sandstone? (p. 331)

___

6. What are the three aspects of sedimentary rocks that help us deduce past climates? (p. 332)

___

7. What role do **ripple marks** play in giving us clues to past events? (p. 333)

___

8. How does **crossbedding** differ from normal stratification? (p. 332)

___

9. Describe **contact metamorphism**. (p. 334)

___

___

10. Why do metamorphic rocks often have a banded appearance? (p. 334)

___

(F) MATCHING

Match the following terms and their meanings:

_______	1. breccia	a. distinct surfaces between strata
_______	2. stratification	b. composite rock made of gravel and pebbles
_______	3. lithification	c. banded appearance of metamorphic rocks
_______	4. shale	d. the layering of rock
_______	5. conglomerate	e. fine grained clastic sedimentary rock
_______	6. foliation	f. the deposition and compaction of rock
_______	7. bedding planes	g. angular fragments in conglomerate rock

(G) ESSAY QUESTIONS

1. How are conglomerates formed?

2. Describe the property of sandstone that makes it economically significant.

3. Classify the following rocks as either igneous, sedimentary, or metamorphic.

basalt ________________

conglomerate ________________

gabbro ________________

gneiss ________________

granite ________________

limestone ________________

marble ________________

quartzite ________________

sandstone ________________

schist ________________

shale ________________

slate ________________

(H) GRAPHICACY

1. Draw an annotated sketch of the rock cycle, using the circle as a guide and arrows to show the direction of cycle processes.

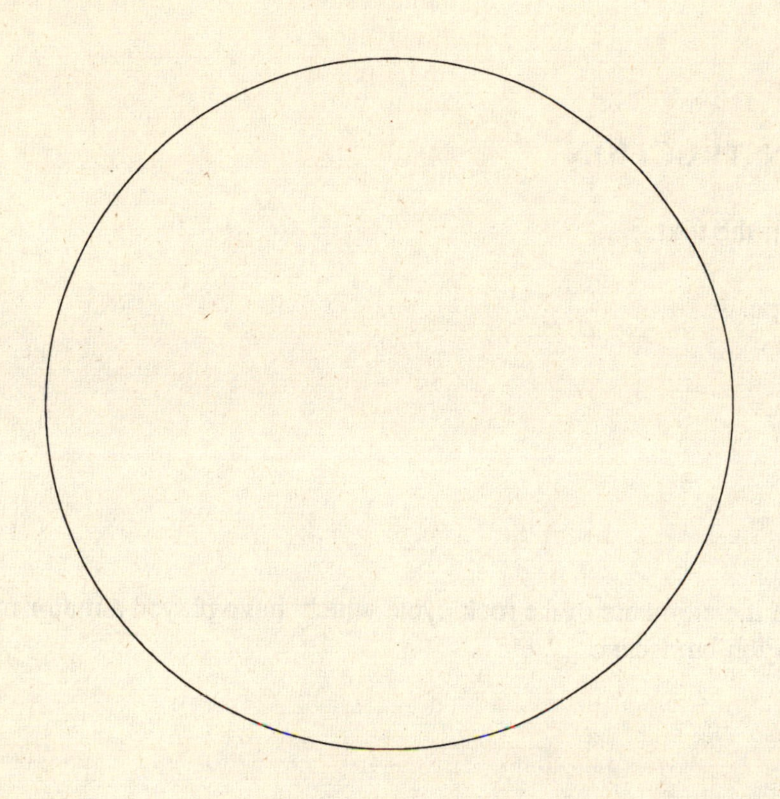

2. Describe the processes that created the landscape in Figure 31-6 in the text.

3. Where in the U.S. could we find rock such as that shown in Figure 31-7? Explain.

(I) HYPOTHESIS CONSTRUCTION

1. Refer to Figure 31-8 in the text.

a. Describe the landscape.

b. Identify and describe the elements of the rock cycle which have played a major role in creating the Grand Canyon landscape.

c. Complete the sentence. The formation of the Grand Canyon is largely the result of

____

____

Unit 32

Plates of the Lithosphere

(A) SUMMARY AND OBJECTIVES

As we deepen our knowledge of the physical geography of the earth, we realize that things are not always as they appear to be. This is particularly true when we examine the nature of the earth beneath our feet. Indeed, the continents themselves have shifted position over the last few hundred years, and are continuing to do so. The "drifting" of the continents is discussed in the broader context of the theory of plate tectonics. Although the idea of continental drift has a long history going back to the seventeenth century, its scientific assessment did not begin until early in this century when Alfred Wegener presented evidence of its occurrence. Since then, the ideas of continental drift and plate movement have been confirmed and greater complexity has been added to the initial concept. An additional theme in this unit is the scientific approach. It often maintains a skeptical view of new scientific ideas and only embraces them once tangible proof has been accumulated.

After studying this unit you should be able to:

1. Understand the origins of the continental drift concept.
2. Explain the nature of continental drift.
3. Identify the different tectonic plates.
4. Describe the various types of plate movement.

(B) KEY TERMS AND OBJECTIVES

Laurasia
lithospheric plates
Pangaea
rifting
Ring of Fire
subduction
tectonic activity
transform faulting

(C) MULTIPLE CHOICE

_____ 1. What is the name of the smaller plate between South America and the larger Pacific Plate? (p. 340)

(a) de Fuca
(b) Nazca
(c) Gorda

_____ 2. What do plates do at the midoceanic ridges? (p. 342)

(a) converge
(b) move laterally
(c) diverge

_____ 3. Which continent was once part of Gondwana? (p. 338)

(a) Asia
(b) South America
(c) North America

_____ 4. What is the average rate of movement of the plates per year? (p. 343)

(a) 1 inch
(b) 6 inches
(c) 2 feet

_____ 5. What do we call a plate boundary where the movement is lateral? (p. 345)

(a) subducting
(b) obverse
(c) transform

(D) TRUE OR FALSE

1. The earth's plates are called lithospheric to show their rigidity and tectonic to show their mobile nature. (p. 339)
T ____ F ____

2. The Atlantic midoceanic ridge is linear and unbroken. (p. 342)
T ____ F _____

3. The earth's crust was formed 250 million years ago from the remains of lava flows. (p. 342)
T _____ F _____

4. In subduction, the plate being forced down is heated and the rock then melts. (p. 343)
T ____ F _____

5. A good example of an oceanic-continental plate convergence is the contact between the Eurasian and Australian plates. (p. 344)
T _____ F _____

(E) SHORT-ANSWER QUESTIONS

1. What was the cartographic insight that Francis Bacon presented in 1619? (p. 337)

____

2. Define **continental drift**. (p. 339)

____

____

3. Why was Wegener's idea not immediately acceptable to many earth scientists? (p. 339)

____

4. What was Arthur Holmes' contribution to the concept of continental drift? (p. 339)

____

5. Where could most of the evidence supporting continental drift be found? (p. 339)

____

6. What role did ocean floor rocks play in confirming continental drift? (p. 339)

____

7. What was **Pangaea** and when did it exist? (p. 339)

____

8. What does plate rifting mean? (p. 343)

____

9. Describe continental-continental plate convergence. (p. 344)

____

10. What plates have contact at the San Andreas fault? In what direction do they move? (p. 345)

____

(F) MATCHING

Match the following terms and their meanings:

_______ 1. Gondwana

_______ 2. Nazca Plate

_______ 3. Pangaea

_______ 4. Gorda Plate

_______ 5. seafloor spreading

_______ 6. Laurasia

_______ 7. Ring of Fire

a. plate between the South American Plate and the Pacific Plate

b. northern region of Pangaea

c. movement of ocean crust away from mid-oceanic ridges

d. southern region of Pangaea

e. tectonically active region around the edges of the Pacific Ocean

f. supercontinent of all the landmasses

g. small plate of the coast of Oregon and northern California

(G) ESSAY QUESTIONS

1. Describe Wegener's circumstantial evidence for continental drift.

2. Discuss the early controversy surrounding the concept of continental drift.

3. Distinguish between oceanic-continental and oceanic-oceanic plate convergence.

(H) GRAPHICACY

1. On the map of the northern Atlantic, draw in the islands of Surtsey and Heimaey. Sketch in the mid-oceanic ridge and the directions of plate movement.

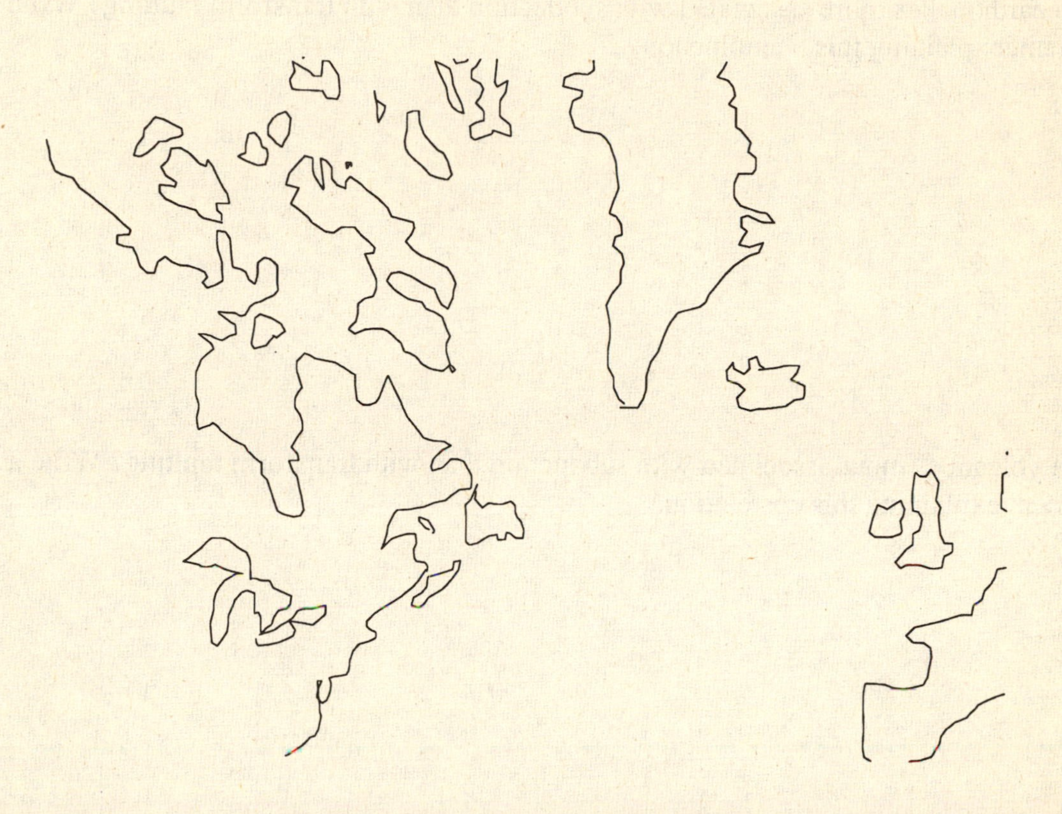

2. Examine the map of plates in Figure 32-3. If the Eurasian plate is moving strongly towards the southeast and the African plate is sliding towards the east, what changes are in store for the Mediterranean Sea and the Red Sea in the next 200 million years?

(I) HYPOTHESIS CONSTRUCTION

1. Compare Figure 32-4, the map showing distribution of earthquakes and volcanoes, with Figure 32-3, the global system of plates.

a. Are earthquakes more associated with subduction than with transform faulting? Write a statement explaining this phenomenon.

b. Are volcanoes more associated with subduction than with transform faulting? Write a statement explaining this correlation.

Unit 33

Plate Movement: Causes and Effects

(A) SUMMARY AND OBJECTIVES

How can building a dam affect the earth's crust? Well, it can change our "solid" earth in ways that are worthy of our consideration. The initial focus in this unit is on the nature of the tectonic plates and on the evolution of our continental landmasses. The scientific approach is highlighted as we examine the alternative scientific explanations for the movement of crustal plates. The authors describe the scientific debate about crustal movement and then focus on the phenomenon of isostasy, defined as the balancing between continental and oceanic areas. This may be the key to understanding the dynamic equilibrium generated in the earth's crust. The unit concludes with the observation that human activity such as dam-building can indeed shift the balance in the earth's crust.

After studying this unit you should be able to:

1. Discuss the mechanisms involved in crustal spreading.
2. Explain how continents evolve.
3. Understand the nature of isostatic movement.
4. Explain the impact of human activities on crustal equilibrium.

(B) KEY TERMS AND CONCEPTS

isostasy
isostatic rebound
rock terrane
suspect terrane
Wrangellia

(C) MULTIPLE CHOICE

_____ 1. What is the approximate age of the earth? (p. 348)
(a) 2 million years
(b) 20.5 million years
(c) 4.5 billion years

_____ 2. What do we call the oldest known rocks? (p. 348)
(a) the Beatles
(b) preJurasic
(c) preCambrian

_____ 3. What type of rocks were created at the margins of shield rocks? (p. 351)
(a) igneous
(b) sedimentary
(c) metamorphic

_____ 4. When does most isostatic uplift occur? (p. 355)
(a) when the siamic root is deep
(b) when the sialic root is deep
(c) when the sialic root is shallow

_____ 5. When did the last of the recent North American icesheets melt? (p. 355)
(a) about 12,000 years ago
(b) about 45,000 years ago
(c) about 5,000 years ago

(D) TRUE OR FALSE

1. Igneous and sedimentary rocks form part of the core areas of the landmasses. (p. 348)
T_____ F_____

2. The interior of the earth has recently been cooling and the rate of internal convection has slowed down. (p. 350)
T_____ F_____

3. When a coalesced landmass becomes subject to crustal spreading, it loses part of its landmass. (p. 351)
T_____ F_____

4. Under mountain ranges, the sialic part of the earth's crust is comparatively thick. (p. 352)
T_____ F_____

5. Rivers erode less spectacularly on the plains than in mountainous regions. (p. 354)
T_____ F_____

(E) SHORT-ANSWER QUESTIONS

1. How has the age of the earth been determined? (p. 349)

____

2. Explain the differences in rock temperature as plates move. (p. 350)

3. What is the major process responsible for plate movement? (p. 350)

4. Define a **suspect terrane**. (p. 351)

5. What is the geologic significance of the area referred to as Wrangellia? (p. 351)

6. Name the world's highest mountain peak? How high is it? (p.352)

7. Define **isostasy**. (p. 353)

8. What is the difference between sialic and siamic rocks? (p. 354)

9. Define **isostatic rebound**. (p. 355)

10. What impact does dam construction have on the crust? (p. 355)

(F) MATCHING

Match the following terms and their meanings:

_______	1. siamic rocks	a. rocks that are similar in age, type, and structure
_______	2. Wrangellia	b. lighter rocks
_______	3. terrane	c. rock masses in the U.S. Pacific Northwest
_______	4. sialic rocks	d. the denser rocks

(G) ESSAY QUESTIONS

1. Describe Harry Hess' explanation for the movement of tectonic plates.

2. Describe the Airy hypothesis.

3. How are isostasy and erosion related?

(H) GRAPHICACY

1. Use the sketch provided as a base and draw a cross-section of the mid-oceanic ridge, illustrating the process of convectional heating currents which move the plates.

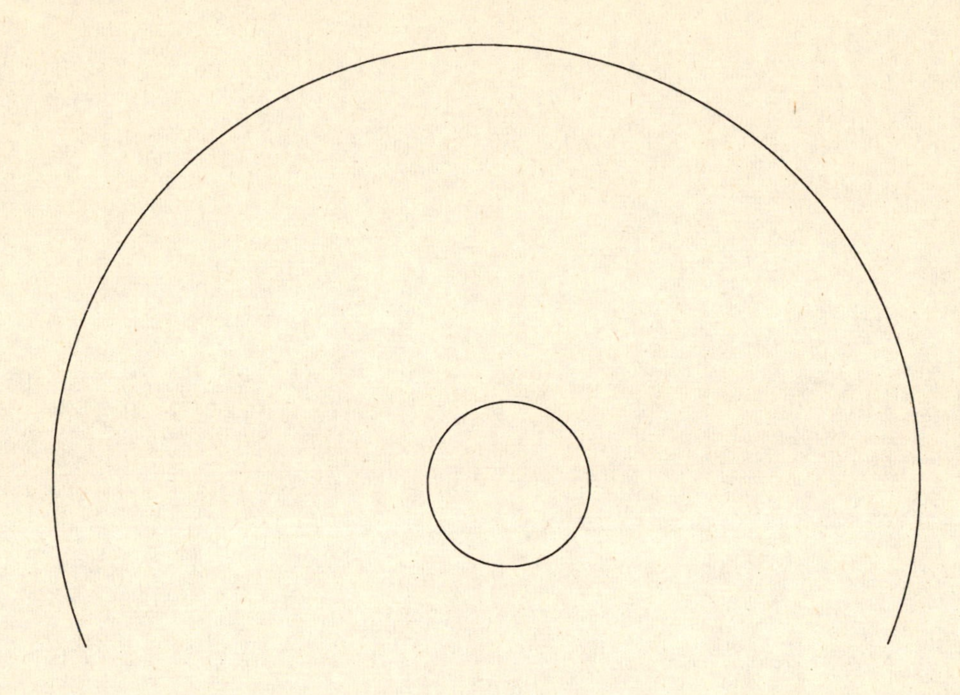

(I) HYPOTHESIS CONSTRUCTION

1. Figure 33-4 shows part of the Himalayan Mountain range.

a. How are these mountains connected to the process of plate movement?

b. Describe the two countervailing forces that will continue to affect the height of the Himalayas?

Unit 34

Volcanism and Its Landforms

(A) SUMMARY AND OBJECTIVES

While our planet provides us with a very habitable environment, it does have its episodes of violent upheaval. Volcanoes are in some places, the most violent of nature's readjustments and they often become part of a folklore and a way of viewing the world. While most volcanoes occur at the boundaries of tectonic plates, the location of some have more obscure explanations that are associated with the little understood phenomenon of hot spots. The discussion in this unit differentiates the various types of volcanoes and the distinctive materials ejected by active volcanoes. The case studies of particularly violent volcanic eruptions help deepen our understanding of the often catastrophic impact which volcanoes have on human populations.

After studying this unit you should be able to:

1. Describe the distribution pattern of volcanoes.
2. Differentiate the various volcanic forms.
3. Explain the impact that volcanism has on human activity.
4. Identify the landscapes created by volcanism.

(B) KEY TERMS AND CONCEPTS

aa
caldera
cinder cones
lahar
nuees ardentes
pahoehoe
pyroclastics

(C) MULTIPLE CHOICE

_____ 1. What percent of the world's volcanoes are located on the seafloor? (p. 357)
(a) 25%
(b) 50%
(c) 75%

_____ 2. What is the best example in the U.S. of a volcanic plateau? (p. 359)
(a) Ohioan
(b) Columbian
(c) Nebraskan

_____ 3. In what country is Mt. Vesuvius located? (p. 367)

(a) Philippines
(b) Chile
(c) Italy

_____ 4. Name the Hawaiian god of volcanism? (p. 362)

(a) Mr. Spock
(b) Pelee
(c) Ardente

_____ 5. Name the largest active shield volcano on the island of Hawaii? (p. 363)

(a) Mauna Loa
(b) Kilauea
(c) Mauna Kea

(D) TRUE OR FALSE

1. Volcanism is mostly associated with either seafloor spreading or with subduction. (p. 358)
T _____ F _____

2. Basaltic lavas are high in silica and low in iron content.
(p. 359)
T _____ F _____

3. The sound of Krakatoa erupting could be heard 3,000 km away. (p. 365)
T _____ F _____

4. Mt. Pinatubo was a major volcanic eruption in Mexico in 1980. (p. 361)
T _____ F _____

5. Lava that hardens and forms a ropy pattern is called pahoehoe. (p. 363)
T _____ F _____

(E) SHORT-ANSWER QUESTIONS

1. Distinguish between an **active** and a **dormant** volcano. (p. 358)

____

____

2. What is meant by an **extinct** volcano? (p. 358)

3. Distinguish between **vents** and **fissure eruptions**. (p. 359)

4. Describe the structure of a composite volcano. (p. 360)

5. Describe a **nuee ardente**. (p. 361)

6. What is a volcanic **hot spot**?(p. 364)

7. Describe a **caldera**. (p. 365)

8. What are the main characteristics of **phreatic eruptions**? (p. 365)

9. What was the global impact of Tambora's 1815 eruption? (p. 366)

10. Name the two towns buried in the Mt. Vesuvius eruption of A.D. 79. (p. 367)

(F) MATCHING

Match the following terms and their meaning:

_______	1. lahars	a. angular, blocky form of lava
_______	2. volcanic bombs	b. a sea wave set off by crustal disturbance
_______	3. pahoehoe	c. material that erupted explosively from a volcano
_______	4. fissure eruptions	d. ropy form of lava
_______	5. aa	e. a mudflow of melted snow and ice created by the heat of a volcano
_______	6. tsunami	f. volcanic eruptions through cracks in the lithosphere
_______	7. pyroclastics	g. masses of lava ejected in a molten state but then solidified in the air

(G) ESSAY QUESTIONS

1. Describe the characteristics of basaltic lava.

2. Where is the Columbia Plateau and how was it formed?

3. What are Nuees Ardentes and how do they affect human populations?

4. What are the differences between a cinder cone and a shield volcano?

5. Where is the Mt. Pinatubo volcano and why is its eruption in 1991 regarded as so significant?

(H) GRAPHICACY

1. On the map of the U.S. northwest, draw in Mt. St. Helens, Mt. Hood, and Mt. Rainier. Also draw in the Gorda, Pacific, and North American Plates and indicate, by arrows, how the eruption of Mt. St. Helens came about.

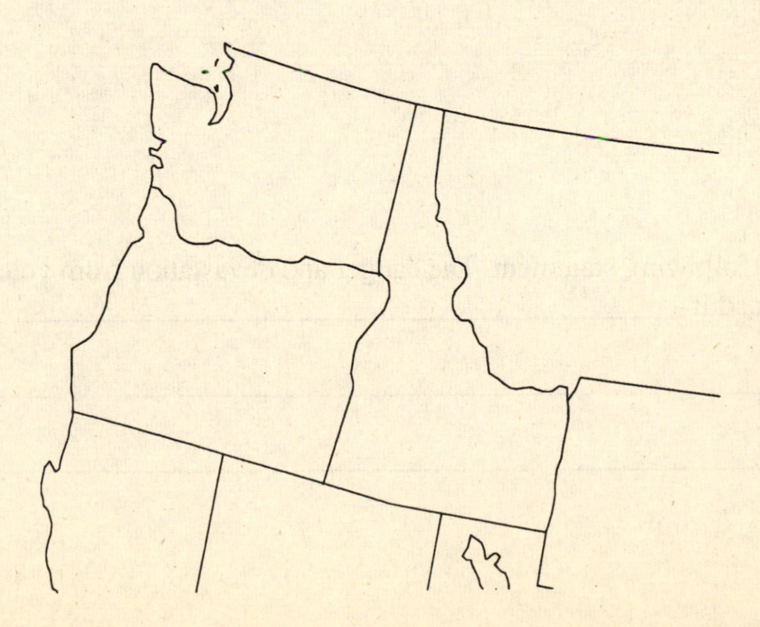

(I) HYPOTHESIS CONSTRUCTION

1. Describe the relationship between hot spots and plate movement.

2. Construct a hypothesis that states this relationship.

3. List the major factors involved in predicting risk from volcanoes.

4. Which of these factors are the most difficult to control?

5. Complete the following statement. The danger and devastation from volcanic eruptions can be drastically reduced if ____

____

____

Unit 35

Earthquakes and Landscapes

(A) SUMMARY AND OBJECTIVES

The 1995 earthquake in Kobe, Japan made us think again about the awesome force that nature can exert at the most unexpected of times. As the world's population grows, the chances of a major earthquake striking such a densely populated area increases very dramatically. Against this backdrop Unit 35 describes earthquakes and the scales for measuring their severity. The notion that "some people live in earthquake zones and some do not" is challenged as recent findings and events indicate that not all earthquakes occur along the edges of lithospheric plates. An example of the unpredictability of earthquakes is relative vulnerability of the eastern part of the U.S. that may be less well-prepared to deal with earthquakes when compared to an area like California. Knowledge and understanding of earthquake mechanisms must be significantly refined so that the impact of earthquakes can be reduced.

After studying this unit you should be able to:

1. Understand the scales for measuring the force of earthquakes.
2. Distinguish the major historical earthquakes.
3. Locate the areas of earthquake activity.
4. Explain the impact of earthquakes on landscapes.

(B) KEY TERMS AND CONCEPTS

epicenter	magnitude
fault	Richter scale
focus	tsunamis

(C) MULTIPLE CHOICE

_____ 1. What term is used to describe the amount of shaking as a quake passes? (p. 370)
(a) velocity
(b) grade
(c) magnitude

_____ 2. What scale measures the intensity of an earthquake? (p. 370)

(a) Mercalli
(b) Richter
(c) Moment Magnitude

_____ 3. Which two plates are responsible for about 80% of all shallow-focus earthquakes? (p. 373)

(a) Gorda and Pacific
(b) Nazca and Pacific
(c) Eurasian and Pacific

_____ 4. Which of these countries is most at risk for serious earthquake damage? (p. 373)

(a) Indonesia
(b) Nigeria
(c) Brazil

_____ 5. What do we call the exposed cliff-like face of a fault line surface? (p. 375)

(a) fault trace
(b) fault scarp
(c) fault plane

(D) TRUE OR FALSE

1. On the Richter scale, a quake of 4 has 10 times as much ground motion as a quake of 3. (p. 370)
T _____ F _____

2. The focus of an earthquake is the point on the surface where the greatest severity is felt. (p. 370)
T _____ F _____

3. The Loma Prieta earthquake of 1989 had a Richter magnitude of 9.5. (p. 372)
T _____ F _____

4. Earthquakes in the midoceanic zone are less severe than those in the Circum-Pacific belt. (p. 373)
T _____ F _____

5. Tsunamis are another name for tidal waves. (p. 376)
T _____ F _____

(E) SHORT-ANSWER QUESTIONS

1. Describe the **Richter Scale**. (p. 370)

2. Compare a Richter scale reading of 6 and one of 8. (p. 370)

3. What advantage does the Mercalli Scale have over the Richter Scale? (p. 370)

4. What is meant by a XI on the Modified Mercalli Scale? (p. 371)

5. Compare the magnitude of the 1906 San Francisco earthquake with that of the 1989 Loma Prieta quake. (p. 372)

6. In addition to shaking ground, what caused most damage in the Anchorage earthquake? (p. 375)

7. What physical factors make earthquakes in the eastern U.S. potentially more devastating than those of the western United States? (p. 374)

8. What social factors make earthquakes in the eastern U.S. potentially more devastating than those of the western United States? (p. 374)

9. Why was the 1985 Mexico City earthquake particularly severe? (p. 358)

10. Explain how tsunamis originate. (p. 359)

____

(F) MATCHING

Match the following terms and their meanings:

_______ 1. fault breccia

_______ 2. focus

_______ 3. fault plane

_______ 4. fault trace

_______ 5. epicenter

_______ 6. fault scarp

_______ 7. magnitude

a. the point on the earth's surface above the rock stress

b. the amount of shaking of the ground

c. the lower edge of a fault scarp

d. the exposed cliff-like face of the fault plane

e. the surface of contact along which blocks of a fault move

f. a band of crushed, jagged rock fragments

g. the place of origin within the earth where the earthquake occurs

(G) ESSAY QUESTIONS

1. Compare and contrast the Richter and Mercalli scales in terms of their objectives and application.

2. Describe the location of the Trans-Eurasian earthquake belt and its relationship to population centers.

3. Contrast the major problems associated with the 1985 Mexico City earthquake and those of the 1964 Anchorage earthquake.

(H) GRAPHICACY

1. Refer to the Earthquake Risk map on page 374 of the textbook. Construct a similar U.S. risk map for volcanoes.

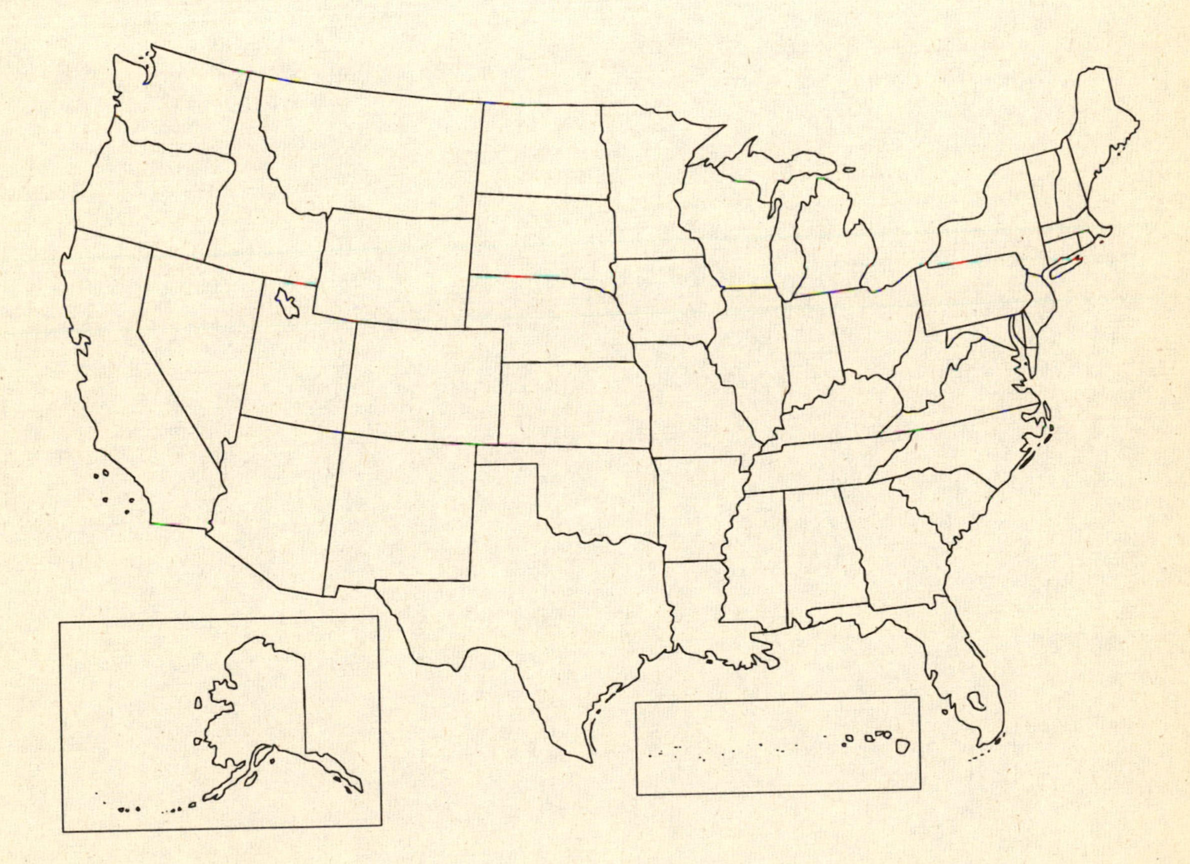

2. Explain the differences in risk patterns on these maps.

(I) HYPOTHESIS CONSTRUCTION

1. Imagine that you are the city planner in Hollister, California. Discuss what planning can be done to limit the damage from future earthquakes.

Unit 36

Surface Expressions of Subsurface Structures

(A) SUMMARY AND OBJECTIVES

Our earth can never be accused of being boring! It is a dynamic planet containing a series of natural processes that change the crust and spatial relationships on the surface. In this unit we are introduced to those crustal stresses that create the surface landform configurations that we see around us. While faulting and folding are the main crustal processes forming and reforming the earth's surface, these initial landforms will eventually succumb to the forces of weathering and gravity to produce the secondary landforms that are most familiar to us. After we classify faults according to the crustal forces at work, we will use the East African rift valley to illustrate the phenomenon of a continually changing earth crust. Our wonderfully complex planet is asking to be explored!

After studying this unit you should be able to:

1. Discuss the elements of rock structure.
2. Understand the formation of various fault structures.
3. Explain the characteristics of folding.
4. Identify and locate the landforms associated with faulting and folding.

(B) KEY TERMS AND CONCEPTS

anticline	folding
dip	horst
epeirogeny	strike
faulting	syncline

(C) MULTIPLE CHOICE

_____ 1. What do we call the compass bearing of a line of intersection between a rock layer and a horizontal plain? (p. 378)

(a) dip
(b) strike
(c) dike

_____ 2. What type of stresses are produced when plates converge and collide? (p. 378)
(a) transverse
(b) tensional
(c) compressional

_____ 3. Echelon faults are also known by what other name? (p. 379)
(a) thrust
(b) parallel
(c) normal

______ 4. What type of fault is represented by the San Andreas fault? (p. 382)
(a) transform
(b) slicken
(c) strike-dip

_____ 5. What term describes the vertical movement of crust with little or no bending of rocks? (p. 385)
(a) recumbent
(b) thrust fault
(c) epeirogeny

(D) TRUE OR FALSE

1. When plates diverge and the crust is spread apart, the stress is known as tensional. (p. 378)
T _____ F _____

2. Another name for a rift valley is a horst. (p. 380)
T _____ F _____

3. When flat-lying sedimentary strata are folded, the cores of the synclines are the younger rocks. (p. 383)
T _____ F _____

4. All types of crustal deformation involve faulting and folding. (p. 384)
T _____ F _____

5. It is possible that an anticline of soft rock can be eroded into a valley. (p. 383)
T _____ F _____

(E) SHORT-ANSWER QUESTIONS

1. How does **dip** differ from strike? (p. 378)

2. How does a **normal fault** differ from a thrust fault? (p. 379)

3. Why is volcanism often associated with rift valleys? (p. 380)

4. What role will the Somali Plate eventually have on Africa's form? (p. 381)

5. What landform does Lake Tanganyika in East Africa represent? (p. 380)

6. Define a **transcurrent** fault. (p. 381)

7. Why are transcurrent faults also known as **strike-slip faults**? (p. 382)

8. What type of rocks are most associated with **folding**? Why? (p. 382)

9. How do the European Alps and the Appalachians differ in their folding? (p. 382)

____

____

10. Distinguish between **recumbent** and **overturned** folding. (p. 383)

____

____

(F) MATCHING

Match the following terms and their meanings:

________	1. fault	a.	a fracture in the crust without displacement
________	2. outcrop	b.	smooth, mirror-like surfaces on a scarp face
________	3. grabens	c.	a block raised between reverse faults
________	4. joints	d.	a series of nearly parallel faults
________	5. slickenslides	e.	a fracture in the crust involving displacement of rock
________	6. echelon	f.	a sunken block between parallel normal faults
________	7. horst	g.	area where exposed rock occurs

(G) ESSAY QUESTIONS

1. Explain the differences between compressional, tensional, and transverse stresses.

2. Describe how a rift valley originates.

3. What is the difference between a primary and a secondary landform?

(H) GRAPHICACY

1. Using Figure 36-1 in the text as a guide, sketch the main features of the landscape illustrated in Figure 36-11. Assume that you are standing in the south of this scene (bottom of page), and draw in the rock layers, the angle of dip, the direction of dip, and the strike.

(I) HYPOTHESIS CONSTRUCTION

1. You are a camera. Examine the unit opening photo on page 377 of the text. Imagine that you are a stationary video camera and that you have been viewing this scene for almost 1 billion years.

a. What changes have you seen in this landscape during the last 600 million years?

b. What tectonic processes could you deduce have been at work here?

c. What evidence would you suggest for these processes?

PART FIVE

SCULPTING THE SURFACE

%%%%%%%%%%%%%%%%%%%%%

Unit 37

The Formation of Landscapes and Landforms

(A) SUMMARY AND OBJECTIVES

While for us time may be thought of as a week, a month, or a semester, in this unit our time framework is broadened to incorporate millions and even billions of years. The geologic span of time is based on scientific research that confirms that the earth is at least 4.5 billion years old. When the earth is placed within this enormous time-frame, we gain a better understanding of the forces of stability and of change within our planet. Through long periods of time the processes of weathering, mass movement, and erosion serve as agents of surface change, operating with differing intensities across the earth and creating landforms such as valleys and mountain ranges. The visible landforms and landscapes result from all of those actions that either wear down or build up the surface over many eons of time. Think of this time context when you feel that 'time is flying'. You may never look at time in quite the same way again.

After studying this unit you should be able to:

1. Explain the processes that lower the earth's relief.
2. Explain the processes that build up the surface.
3. Identify landforms and landscapes of degradation and aggradation.
4. Understand the distinctions in the geologic time chart.

(B) KEY TERMS AND CONCEPTS

aggradation
degradation
erosion
geological time scale
landform
landscape
mass movement
pre-Cambrian

(C) MULTIPLE CHOICE

_____ 1. When a river builds a delta at its mouth, what process is taking place? (p. 390)
(a) mass movement
(b) degradation
(c) aggradation

_____ 2. What Latin phrase refers to a change in the present location, without removal to another site? (p. 390)

(a) pari passu
(b) in situ
(c) passe

_____ 3. Which of these rock types often contain sinkholes? (p. 392)

(a) limestone
(b) sandstone
(c) Green Day

_____ 4. If there were no uplift forces, how long would it take for the continents to be planed sea level? (p. 392)

(a) 50 million years
(b) 120 million years
(c) 270 million years

_____ 5. Which era is located in the Tertiary period of geologic time? (pp. 393)

(a) Paleozoic
(b) Cenozoic
(c) Mesozoic

(D) TRUE OR FALSE

1. Only those areas that experience subduction and other forms of deformation are subjected to vertical movement that affect the landscape. (p. 389)
T _____ F _____

2. A river valley would be defined as a landform rather than as a landscape. (p. 389)
T _____ F _____

3. Stream erosion involves both transportation and breakdown of rocks. (p. 391)
T _____ F _____

4. Erosion is limited by the amount of space on the ocean floors. (p. 392)
T _____ F _____

5. In the geologic time scale, epochs refer to the evolution of all life on earth. (p. 393)
T _____ F _____

(E) SHORT-ANSWER QUESTIONS

1. How do secondary landforms differ from primary ones? (p. 388)

____

____

2. Distinguish between **landforms** and **landscapes**. (p. 373)

____

____

3. Describe the process of **mass movement**. (p. 390)

____

____

4. What occurs during the process of **erosion**? (p. 390)

____

____

5. Name the three main factors influencing the ability of streams to erode. (p. 391)

____

6. What is the significance of the Precambrian era for geomorphologists? (p. 392)

____

____

7. Describe the process of **isostasy**. (p. 392)

____

____

8. Describe **Quaternary** rocks. (p. 393)

____

____

9. Distinguish between eras, periods, and epochs. (p. 393)

____

____

10. Name two examples of physiographic regions. (p. 394)

____

(F) MATCHING

Match the following terms and their meanings:

_______	1. tributaries	a. an era of recent life starting about 65 million years ago
_______	2. landscapes	b. the wearing down of landmasses
_______	3. degradation	c. rocks older than 57 million years
_______	4. Cenozoic	d. an aggregation of landforms
_______	5. landforms	e. smaller streams that flow into the main stream
_______	6. regolith	f a single typical unit in the landscape
_______	7. Precambrian	g. a layer of weathered rock overlying the bedrock

(G) ESSAY QUESTIONS

1. How does gravity affect the landform change process?

2. Describe what typically happens to a rock during the process of erosion.

(H) GRAPHICACY

1. On the map of China, locate the following:
   - the Chang Jiang River illustrated in Figure 37-2.
   - the city of Guilin mentioned in the opening photo caption on page 388 of the text.

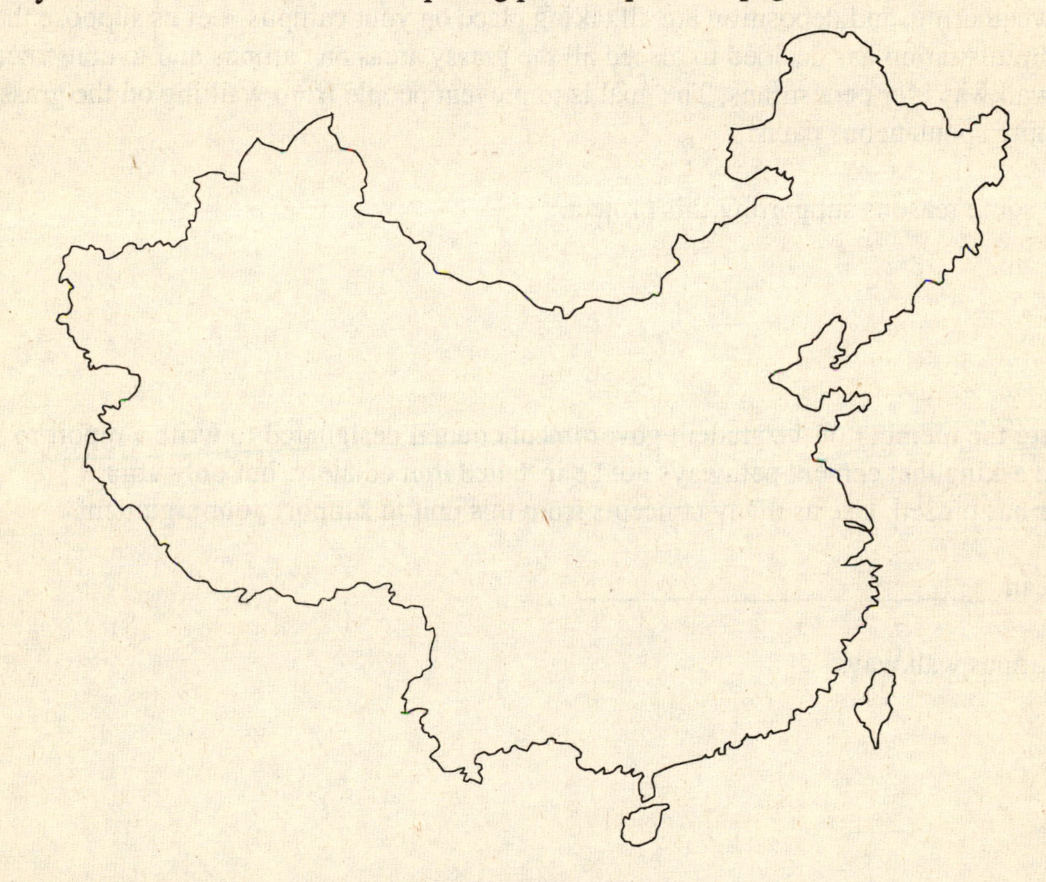

a. Draw in the Chang Jiang and the Three Gorges section.

b. What are the benefits from building a dam in this region?

c. Discuss the environmental damage which a large scale dam could do.

(I) HYPOTHESIS CONSTRUCTION

Erosion, weathering, and deposition are all taking place on your campus. Let us suppose that your college administration has decided to reseed all the grassy areas on campus and to construct concrete walkways for pedestrians. The goal is to prevent people from walking on the grass and from forming spontaneous paths.

a. Name some reasons supporting this project.

b. You are the member of the student government council designated to write a report to the Dean asking that cement pathways not be installed immediately, but only after a semester has passed. Use as many concepts from this unit to support your argument.

To: Dean ________________________

Re: Campus walkways

Unit 38

Weathering Processes

(A) SUMMARY AND OBJECTIVES

In the movie "The Grapes of Wrath", we are reminded of the unforgiving power of nature as large numbers of people were forced to leave a blighted area. It is a classical tale of how the physical geography of a place can become a "character" in the very story of human existence. The story also reminds us that if we lack a good understanding of how the physical world works, then we may be doomed to make the same bad ecological decisions. We need to understand the processes of rock breakdown by mechanical, chemical, and biological weathering so that we can prevent soil degradation and increase our ability to grow food. The physical -human continuum referred to in Unit 1 is evident here as well. We see that occurrences of such ecological disasters are almost always a combination of physical and cultural actions. Disciplines, like geography, that consciously incorporate both elements can surely help us not only survive but also prosper.

After studying this unit you should be able to:

1. Differentiate the various types of weathering.
2. Understand the processes involved in weathering.
3. Relate weathering to climatic zones.
4. Identify the role of human activity in hastening the weathering process.

(B) KEY TERMS AND CONCEPTS

carbonation	oxidation
felsenmeer	rock sea
frost action	scree
hydrolysis	talus cone

(C) MULTIPLE CHOICE

_____ 1. What is another term for a talus cone? (p. 397)
(a) wedging
(b) scree slope
(c) felsenmeer

_____ 2. What type of weathering is present when the limestone in calcium carbonate dissolves and leads to rock disintegration? (p. 397)

(a) mechanical
(b) biological
(c) chemical

_____ 3. What name is given to the geometrical structures of stone found in the soil of the Arctic region? (p. 397)

(a) stone nets
(b) felsenmeer
(c) Sassy

_____ 4. What name is given to the flaking off of the outer shells of granite boulders? (p. 398)

(a) spheroidal weathering
(b) hydrolotic weathering
(c) biological weathering

_____ 5. What type of map would tell us a most about the nature of weathering in different parts of the world? (p. 400)

(a) vegetation
(b) climate
(c) landuse

(D) TRUE OR FALSE

1. Water increases in volume by about 9% when it freezes. (p. 396)
T _____ F _____

2. Mechanical weathering involving expansion of crystals in rock crevases only occurs in moist regions. (p. 397)
T _____ F _____

3. Chemical weathering is the least effective agent of rock destruction in humid areas. (p. 398)
T _____ F _____

4. The sole reason for the 1930s Dust Bowl was the severe drought that lasted 6 years. (p. 399)
T _____ F _____

5. Soils are thicker in equatorial areas than in polar areas. (p. 400)
T _____ F _____

(E) SHORT-ANSWER QUESTIONS

1. Describe the process of **mechanical weathering.** (p. 396)

____

____

2. Why is frost action damaging to rocks? (p. 396)

____

____

3. Describe the type of mechanical weathering that occurs in arid climates. (p. 397)

____

____

4. Why is hydrolysis so effective in weathering? (p. 398)

____

____

5. What are the products of oxidation? (p. 398)

____

6. In what climatic region is oxidation most common? Why? (p. 398)

____

7. In what climatic zone does carbonation mostly occur? Why? (p. 398)

____

8. What type of weathering is most effective in humid areas? Why? (p. 398)

____

9. Describe **biological weathering.** (p. 398)

____

____

10. What type of weathering is predominant in the *BW* climate regions? Why? (p. 400)

____

____

(F) MATCHING

Match the following terms and their meanings:

_______	1. talus cone	a. a process of rock decomposition when water combines with minerals
_______	2. felsenmeer	b. chemical weathering when rocks react with oxygen
_______	3. hydrolysis	c. cone-shaped rock debris at the base of a steep slope
_______	4. oxidation	d. pieces of rock accumulated near their original location

(G) ESSAY QUESTIONS

1. What are stone nets and what do they tell us about the weathering process?

2. Describe the three main ways in which human activity contributes to weathering.

(H) GRAPHICACY

1. Refer to Figure 38-7 in the text.
   Describe how this open-pit mine is contributing to weathering.

2. Refer to Figure 38-4.
   a. What is your estimation of the angle of this talus slope?

   b. Why does the angle lie within talus limits?

3. Note examples of the following weathering processes on your campus:
   a. Mechanical

   b. Chemical

c. Biological

(I) HYPOTHESIS CONSTRUCTION

1. Refer to the Perspective Box on the Dust Bowl on page 399 in the textbook.
   a. List the physical factors that created the Dust Bowl.

   b. List the human actions that contributed to the Dust Bowl.

   c. Write a short editorial statement (6 lines) that begins: Future Dust Bowls can best be prevented if

____

____

____

____

____

____

Unit 39

Mass Movements

(A) SUMMARY AND OBJECTIVES

An important task for people working in the field of *applied geography*, is evaluating landscapes and informing people about the conditions of the physical world. While humans have wide choices in where to live, residential areas are often located in dangerous topography - landscapes can and do change under the force of gravity. These mass movements of earth and soil can be detrimental to human settlements and have to be studied before settlement occurs. Slopes that may seem harmless can become sliding masses of rock if the elements of water and gravitational pull are present. Factors such as climate, soil type, and vegetation determine the speed and extent of mass movement and also the level of danger for people. Most of the contributing factors are naturally occurring but can be exacerbated by human actions such as slope adjustment and dam-building. Detailed knowledge of mass movement is important because it allows for better land-use planning and management and can potentially avoid disaster.

After studying this unit you should be able to:

1. Understand the role of gravity in moving earth material.
2. Explain the nature of mass movement.
3. Distinguish the various types of mass movement.
4. Discuss the importance of mass movements for human activity.

(B) KEY TERMS AND CONCEPTS

angle of repose
earth flow
mass wasting
mudflow
slumping
solifluction

(C) MULTIPLE CHOICE

_____ 1. Which surface would have the highest angle of repose? (p. 402)
(a) granite
(b) loose rocks
(c) loose sand

_____ 2. What are the main ingredients for flow movements to occur? (p. 404)

(a) oil and water
(b) oil and gravity
(c) water and gravity

_____ 3. What term describes soil and rock that is saturated with water and flows as one mass? (p. 404)

(a) slumping
(b) moshing
(c) rockslide

_____ 4. Which movement has the highest water content? (p. 407)

(a) mudflow
(b) rock fall
(c) creep

_____ 5. Which movement has the highest rate of speed? (p. 407)

(a) creep
(b) slide
(c) rock fall

(D) TRUE OR FALSE

1. The steeper the slope, the stronger the shearing stress of the material. (p. 402)
T _____ F _____

2. If mass movement did not occur, the entire gradational system would be slowed down. (p. 403)
T _____ F _____

3. Solifluction is most common in the tropical zones. (p. 404)
T _____ F _____

4. The difference between landslides and another mass movements is the amount of load being transported. (p. 405)
T _____ F _____

5. Mudflow involves major sections of regolith, soil, or weakened bedrock. (p. 405)
T _____ F _____

(E) SHORT-ANSWER QUESTIONS

1. What are the two main factors that determine the rate of rock movement on slopes? (p. 403)

____

2. Define the **angle of repose**. (p. 402)

____

3. What happens to the angle of repose during stream erosion? (p. 402)

____

____

4. Distinguish between **creep** and **flow** movements. (p. 403)

____

____

5. What clues are there in the landscape that creep has occurred? (p. 403)

____

____

6. How does vegetation affect soil creep? (p. 403)

____

7. Describe an **earth flow**. (p. 404)

____

____

8. How do landslides differ from other flow movements? (p. 405)

____

____

9. Name four factors that can cause landslides. (p. 405)

____

____

10. How do landslides differ from creep? (p. 405)

____

____

(F) MATCHING

Match the following terms and their meanings:

_______	1. mudflow	a. the downslope pull of material
_______	2. solifluction	b. permanently frozen subsoil
_______	3. shearing stress	c. special form of creep in tundra areas where the saturated soil has thawed
_______	4. permafrost	d. stream of fluid, lubricated mud

(G) ESSAY QUESTIONS

1. Describe the role of water in downslope movement.

2. How is mass movement related to the process of weathering?

3. Compare the impacts on the landscape from soil creep and flow movements.

(H) GRAPHICACY

1. Refer to the map of U.S. landslide susceptibility on page 406 in the text. Discuss the probable reasons for landslides in three different regions of the United States.

a. The Appalachian region

b. the lower Mississippi valley

c. North and South Dakota

(I) HYPOTHESIS CONSTRUCTION

1. Refer to Figure 39-13.

   a. Based on the information in this sketch, describe the differences between mudflow and creep.

   b. Write two hypothesis sentences that addresses these differences.

(i) The higher ____ of mudflow

causes it to ____

(ii) The lower ____ of creep

causes it to ____

Unit 40

Water in the Lithosphere

(A) SUMMARY AND OBJECTIVES

"Water, water everywhere, and not a drop to drink" is an apt phrase when you are out on the ocean because that water is surely not drinkable. But our concern here is with water we use for human consumption - water found either in rivers, lakes, or under the surface of the ground. Again we need our earth science knowledge to help determine where this water is located and how it can be extracted for human use. Factors such as infiltration, permeability, and runoff are introduced as critical variables in effective water management. With a rapid growth in global population, it becomes imperative that reliable, clean water supplies be available for all. Beyond the fundamentals of stream channels and the measurement of stream flow we need a broader understanding of the often negative impact that humans can have on underground water supplies. Unless we take special care, water supplies in both the developed and the less developed world may, in the future, become even more critical sources of conflict. As stewards of this planet, we are perhaps ultimately judged in the extent to which we maintain the essential requirement for human existence - a clean supply of water.

After studying this unit you should be able to:

1. Understand the movement of water in the soil.
2. Explain the characteristics of river channels.
3. Distinguish the various types of springs.
4. Differentiate the many uses of rain.

(B) KEY TERMS AND CONCEPTS

aquiclude	infiltration
aquifers	interception
gradient	runoff
groundwater	velocity

(C) MULTIPLE CHOICE

_____ 1. Which trees have the lowest rates of interception? (p. 410)
(a) fir
(b) hemlock
(c) eucalyptus

_____ 2. What is the average rate of heavy rainfall? (p. 410)
(a) 3 mm per hour
(b) 5 mm per hour
(c) 8 mm per hour

_____ 3. What percent of the world's fresh water is found in rivers? (p. 412)
(a) .03%
(b) 1.5%
(c) 5.5%

_____ 4. What percent of the total freshwater supply is located under the ground? (p. 412)
(a) 12%
(b) 25%
(c) 40%

_____ 5. What term describes the downward movement of water through the pores and spaces in the soil, under the force of gravity? (p. 413)
(a) field capacity
(b) capillary action
(c) percolation

(D) TRUE OR FALSE

1. The largest runoff occurs at the base of the slope, just before the runoff enters the river. (p. 411)
T _____ F _____

2. The Amazon has a lower gradient than the Mississippi River. (p. 411)
T _____ F _____

3. Streams tend to move fastest along the sides of the channel. (p. 411)
T _____ F _____

4. Some mountain torrents may have lower velocities than the placid lower Mississippi. (p. 412)
T _____ F _____

5. Mudstone and shale make very good aquifers. (p. 413)
T _____ F _____

(E) SHORT-ANSWER QUESTIONS

1. What are **hydraulic societies**? (p. 409)

2. Name four factors that determine the rate of infiltration. (p. 410)

3. Distinguish between impermeable and permeable surfaces. (p. 410)

4. What is the relationship between gradient and stream velocity? (p. 411)

5. Define the **discharge** of a river. (p. 411)

6. Name two purposes where information of the stream discharge is used. (p. 412)

7. Distinguish between the **zone of aeration** and the **zone of saturation**. (p. 412)

8. Why is soil moisture rarely at zero? (p. 413)

9. What is the difference between a confined and an unconfined aquifer? (p. 414)

____

____

10. How are **springs** formed? (p. 414)

____

____

(F) MATCHING

Match the following terms and their meanings:

	Term		Meaning
_______	1. gradient	a.	the amount of rainfall that falls in a given time
_______	2. interception	b.	the downward movement of water through pores in the soil
_______	3. hydrograph	c.	circular movement in streams
_______	4. intensity of rainfall	d.	amount of the drop in the local water table
_______	5. eddies	e.	the difference in elevation between two points
_______	6. percolation	f.	the effect of vegetation preventing rain from reaching the ground
_______	7. drawdown	g	a record of a river's discharge over time

(G) ESSAY QUESTIONS

1. Name and describe the five factors that influence infiltration.

(H) GRAPHICACY

1. Refer to Figure 40-2 in the text.
   a. Explain what factors determine the different infiltration rate curves for meadows and cornfields illustrated in this graph.

   b. Draw a similar graph to indicate differential infiltration rates for savanna and desert landscapes.

minutes

   c. Explain the nature of these different graphed infiltration rates.

(I) HYPOTHESIS CONSTRUCTION

1. Complete the sentences.

The ________________ the runoff rate, the ________________________ the infiltration.

The ________________ the infiltration, the________________________ the runoff rate.

2. Draw a line graph showing this relationship; place the infiltration rate on the horizontal axis.

Slopes and Streams

(A) SUMMARY AND OBJECTIVES

What a thrill it must have been to be one of the explorers far in the interior of China, often following false leads and then finally finding the small, seemingly insignificant and puny rivulet that would eventually become the mighty Mekong River! If physical geography can best be studied by "the soles of your shoes" then this was surely an experience not to be missed. Rivers have always played a major role in human existence, serving as a place for settlement, for plant and animal domestication, providing a source of water and a mode of transportation. In this unit we are introduced to this wonderful world of rivers and we can, at least in our minds, trace the rivers from their origins in the highlands, down to where they enter the ocean. Streams should be seen as part a system consisting of main streams, tributaries, valleys, and drainage basins. Intensive human activity in the vicinity of streams often increases the risk of hazard, particularly, as we are reminded, on floodplains where perceived short term solutions to containing the water may lead to greater danger in the future.

After studying this unit you should be able to:

1. Explain the processes involved in stream erosion.
2. Discuss the relationship between streams and their drainage basins.
3. Understand the dangers of floodplain settlement.
4. Interpret the stream as a system.

(B) KEY TERMS AND CONCEPTS

abrasion
alluvium
base level
corrosion

floodplain
rills
sheet erosion
suspension

(C) MULTIPLE CHOICE

_____ 1. What is the name given for the thin layer of water that moves downslope without being confined to channels? (p. 419)

(a) tributary
(b) sheet erosion
(c) sheet flow

_____ 2. What is the term for the initial small channels that eventual merge to form streams? (p. 419)

(a) rills
(b) trunks
(c) runs

_____ 3. Which two rivers are the main tributaries of the Mississippi River? (p. 420)

(a) the Hudson and the Missouri
(b) the Georgian and the Kansan
(c) the Missouri and the Ohio

_____ 4. What term refers to the erosive action of boulders, pebbles, and smaller grains of sediment as they are carried along the river valley? (p. 422)

(a) traction
(b) abrasion
(c) corrosion

_____ 5. What do we call the process where sand and gravel fragments are lifted by the speed of the water and bounced along on the streambed? (p. 423)

(a) capacity
(b) suspension
(c) saltation

(D) TRUE OR FALSE

1. Drainage basins are separated by topographic barriers called watersheds. (p. 420)
T _____ F _____

2. The source of a river valley is extended upslope in a process called backward erosion. (p. 422)
T _____ F _____

3. In terms of the volume of rock removed, corrosion is the least important form of erosion. (p. 422)
T _____ F _____

4. A river's base level could lie a few meters below sea level. (p. 424)
T _____ F _____

5. A graded stream represents a balance among valley profile, water volume, velocity, and transported load. (p. 426)
T _____ F _____

(E) SHORT-ANSWER QUESTIONS

1. Describe the process of **denudation**. (p. 419)

2. How can sheet erosion cause considerable erosion? (p. 419)

3. Why does sediment yield not reveal all erosion at work in the stream? (p. 420)

4. What does the valley shape reveal about stream activity? (p. 421)

5. How does **abrasion** deepen and widen the stream channel and valley? (p. 422)

6. Distinguish between **suspension** and **saltation**. (p. 422)

7. How does **stream competence** differ from stream capacity? (p.423)

8. How are floodplains built up? (p. 424)

____

9. Describe the difference between **absolute** and **local** base level. (p. 424)

____

____

10. Describe a stream that is **graded**. (p. 426)

____

____

(F) MATCHING

Match the following terms and their meanings:

Term	Meaning
_______ 1. rills	a. material deposited by streams
_______ 2. sheet flow	b. moving of particles in a stream through a series of jumps
_______ 3. fluvial	c. small channels of water
_______ 4. alluvium	d. low-lying ground adjacent to the stream channel
_______ 5. saltation	e. rate at which soil absorbs water from the surface
_______ 6. infiltration	f. referring to running water capacity
_______ 7. floodplain	g. rain that is not absorbed by the soil and runs off

(G) ESSAY QUESTIONS

1. What is a drainage basin?

2. Why is it important for earth scientists to speak of drainage basins rather than isolated streams?

(H) GRAPHICACY

1. Refer to Figure 41-4.

a. Describe what has happened to the interfluvial areas (areas between the streams) from time *a* to time *b*.

b. How has tributary formation changed from *a* to *b* and *c*?

c. Draw a sketch to indicate the valley cross-sections in the middle stage.

(I) HYPOTHESIS CONSTRUCTION

1. A hypothesis often includes relationships between many factors. One way to represent this set of relationships is to systematically isolate them. In unit 1 we presented an organizational formula used in scientific reasoning.

The Dependent variable (DV) is determined by an Independent Variable (IV) + another Independent Variable (IV) OR in shorthand fashion

$$DV = IV + IV + IV$$

The *dependent variable* is the one we wish to analyze, in this case the level of stream grading. It *depends* on other factors. *Independent variables* are those that determine the dependent variable. In our example they include factors such as the width and depth of the channel.

a. Write out a hypothesis statement with level of stream grading as the dependent variable. Use the shorthand method first.

b. Now construct a hypothesis statement that includes these factors.

Unit 42

Degradational Landforms of Stream Erosion

(A) SUMMARY AND OBJECTIVES

While our main interest lies with the forces that carve up and erode the surface landscape, we also need to note the physical structures within which that erosion occurs. This unit examines the impact of geologic, subsurface structures on the development of stream systems. Many of the landforms we see around us reflect the erosive force of streams as they work within the limits of the earth structures to create broad landscape change. Of all of our geologic forces, stream systems are especially susceptible to rock structure and climate conditions as they create new landscapes. This unit concludes with a discussion of the contribution of a prominent geographer, William Morris Davis whose theories are still studied in the continual analysis of the changing landscape.

After studying this unit you should be able to:

1. Explain the role of geological structure on stream erosion.
2. Differentiate the features of different drainage systems.
3. Appreciate William Morris Davis' contribution to geography.
4. Understand the nature of the debate on landscape formation.

(B) KEY TERMS AND CONCEPTS

butte
cuesta
drainage density
hogback
mesas
peneplain
stream piracy
water gap

(C) MULTIPLE CHOICE

_____ 1. What term describes a landform that has a low fairly steep ridge on one side and a very gentle slope on the other? (p. 430)

(a) cuesta
(b) hogback
(c) dike

_____ 2. Which of these is an example of a granite dome? (p. 432)

(a) Ship Rock
(b) Trump Towers
(c) Stone Mountain

_____ 3. What type of drainage patterns are mostly present in the dome structure of South Dakota's Black Hills? (p. 435)

(a) radial
(b) dendritic
(c) annular

_____ 4. What type of drainage patterns would be found in glaciated landscapes? (p. 436)

(a) trellis
(b) deranged
(c) contorted

_____ 5. What is the name given to the most prominent not-yet-eroded remnants on a peneplain? (p. 439)

(a) monadnock
(b) inselberg
(c) butte

(D) TRUE OR FALSE

1. Buttes are mesas that have experienced more erosion. (p. 433)
T _____ F _____

2. Normally, the higher the drainage density, the greater the erosional efficiency of the stream system. (p. 435)
T _____ F _____

3. The most common drainage pattern is the trellis type. (p. 436)
T _____ F _____

4. A superimposed stream maintains its course regardless of the changing rock structures encountered. (p. 436)
T _____ F _____

5. Some slopes are worn down backward rather than downward as Davis insisted. (p. 439)
T _____ F _____

(E) SHORT-ANSWER QUESTIONS

1. How does stream erosion create a **hogback**? (p. 430)

____

____

2. How does the Black Hills topography influence the area's **human** geographic features? (p. 431)

____

____

3. What are the two ways in which bedrock can influence landscape development? (p. 433)

____

____

4. What does stream rejuvenation mean? (p. 433)

____

____

5. Define **drainage density**. (p. 435)

____

____

6. What are the three categories used to describe drainage density? (p. 435)

____

7. What is the relationship between drainage density and erosion? (p. 435)

____

____

8. Describe a **radial** drainage pattern. (p. 435)

____

9. Under what conditions does a **trellis** drainage pattern develop? (p. 436)

____

10. How do **superimposed** and **antecedent** streams differ? (p. 437)

____

____

(F) MATCHING

Match the following terms and their meanings:

	Term		Meaning
_______	1. annular drainage	a.	the study of bedrock
_______	2. water gap	b.	the "capture" of a segment of one stream by another
_______	3. hogback	c.	drainage system that develops on domes
_______	4. centripetal drainage	d.	a remnant of resistant rock rising above a peneplain
_______	5. stream piracy	e.	a prominent, steep-sided ridge
_______	6. monadnocks	f.	a pass in a ridge through which a stream flows
_______	7. lithology	g.	streams flowing into a central basin

(G) ESSAY QUESTIONS

1. On page 430 of the text, the authors distinguish between a *geological* and a *geographical* phenomenon. Explain this distinction.

2. Describe how the peneplain in the Davis model develops.

3. Why was the pediplane later proposed as an alternative explanation to peneplain?

(H) GRAPHICACY

1. What is the dip of the dike in Figure 42-1a?

2. Describe the difference between a mesa and a butte as illustrated in Figure 42-6.

(I) HYPOTHESIS CONSTRUCTION

1. Dr. Martin Luther King Jr's famous 1963 "I have a dream" speech contained many physical geographic references:

"I have a dream
Let freedom ring
From the prodigious hilltops of New Hampshire
Let freedom ring
From the mighty mountains of New York
Let freedom ring
From the heightening Alleghenies of Pennsylvania
Let freedom ring
From the snow-capped Rockies of Colorado
Let freedom ring
From the curvaceous slopes of California
But not only that,
Let freedom ring
From Stone Mountain of Georgia
Let freedom ring
From Lookout Mountain of Tennessee
And when we let it ring
All of God's children will be able to join hands
Free at last. "

a. List four landform features in this part of the speech and the processes that created them.

b. Locate your four features on a U.S. map.

c. Why do think Dr. King incorporated these image references in his speech?

Unit 43

Aggradational Landforms of Stream Erosion

(A) SUMMARY AND OBJECTIVES

Those indelible images of "The Great U.S. Midwest Flood of 1994" remind us of the powerful, unstoppable force that a river can be. The term floodplain became all too literal when both the Mississippi and the Missouri rivers overflowed their banks and inundated towns and farms. Because of the low relief of river valleys in their late stage we can see how this could easily happen. For nature, this is a normal process, but for humans the floodplain is filled with dilemmas. While the floodplain contains those valuable fertile alluvial soils so important for agricultural development, under certain weather conditions, it can become a region of danger and great loss. Our fascination with the Mississippi and the Nile is in part due to the important roles that these rivers have played in economic and social history. They represent, in different settings, the great complexity of human-land relationships.

After studying this unit you should be able to:

1. Explain the nature of alluvial fans.
2. Describe the features of a floodplain.
3. Distinguish the various types of deltas.
4. Interpret the basics of remote sensing.

(B) KEY TERMS AND CONCEPTS

alluvial fan	distributary
bajada	levee
braided stream	meander
delta	oxbow lake

(C) MULTIPLE CHOICE

_____ 1. What do we call a stream that only flows intermittently? (p. 442)

(a) ephemeral
(b) consistent
(b) laid back

_____ 2. What term describes the intertwining of many stream channels into one river? (p. 442)

(a) roped stream
(b) braided stream
(c) bound stream

_____ 3. What name is given to the smooth, gently sloping bedrock surface that extends outward from the foot of the highlands? (p. 443)

(a) apron
(b) pavement
(c) piedmont

_____ 4. What name is given to the landforms that bound the floodplain? (p. 445)

(a) belts
(b) bluffs
(c) ridges

_____ 5. What Spanish word best describes the coalescing of alluvial fans? (p. 443)

(a) bajada
(b) encada
(c) la strada

(D) TRUE OR FALSE

1. In alluvial fans, the thickest stratified layers are found towards the mountain front and then thin out progressively. (p. 443)
T _____ F _____

2. Erosion on the inside of stream meanders is greater than on the outside. (p. 443)
T _____ F _____

3. As stream capacity increases, the river stops degrading and starts building landforms. (p. 443)
T _____ F _____

4. Distributaries flow into the main stream. (p. 448)
T _____ F _____

5. Both the Mississippi and the Zaire rivers have large deltas. (p. 448)
T _____ F _____

(E) SHORT-ANSWER QUESTIONS

1. How does the amount of streamflow affect alluvial fan formation? (p. 442)

____

____

2. How do alluvial fans help human occupancy of the land? (p. 443)

____

____

3. Describe the two movements of stream meanders. (p. 426)

____

____

4. Define a **floodplain**. (p. 444)

____

____

5. Why does the channel of the floodplain become an ever smaller part of the valley? (p. 444)

____

____

6. What is the function of a levee? (p. 445)

____

7. Why are some river terraces not paired? (p. 447)

____

____

8. How do **incised meanders** get formed? (p. 447)

____

____

9. Under what conditions does a **birdfoot delta** develop? (p. 448)

____

10. How does isostasy affect delta formation? (p. 448)

____

____

(F) MATCHING

Match the following terms and their meanings:

_______	1. bajada	a.	bends in the river channel
_______	2. braided stream	b.	stream that flows intermittently
_______	3. terraces	c.	an assemblage of continuous alluvial fans
_______	4. ephemeral stream	d.	smooth, gently sloping bedrock under an alluvial cover
_______	5. meanders	e.	crescent-shaped lakes formed when a meander is cut off
_______	6. pediment	f.	a stream that divides into smaller intertwined channels
_______	7. oxbow lakes	g.	remnant levels of the older floodplain

(G) ESSAY QUESTIONS

1. Describe three conditions under which alluvial fans develop.

2. Explain the main factors determining formation of deltas.

(H) GRAPHICACY

1. Describe the features of the bluffs as illustrated in Figure 43-5.

2. Draw three sequential sketches to show how an oxbow lake can develop.

(I) HYPOTHESIS CONSTRUCTION

1. Compare the remote sensing image of the Nile Delta in Figure 43-13 with the astronaut's photograph in Figure 41-5.

   a. Determine which areas in the remote sensing image are cultivated, populated, or covered with water.

   b. What conclusions can you draw from the color usage in remote sensing imagery?

Unit 44

Karst Processes and Landforms

(A) SUMMARY AND OBJECTIVES

Beyond its extensive conflict, the region referred to as ex-Yugoslavia is known for many intriguing and beautiful landforms and landscapes, many of them below the surface. The caves, towers, and sinkholes in Croatia and Slovenia represent the outcome of numerous percolating processes found in limestone regions extending throughout the region. The stalagtites and stalagmites that abound in these karst caves are indicative of landforms that are constantly being formed and changed by water as it interacts with soluble limestone. Karst, the name for these landscapes, is indigenous to the Balkans and they remind us of the natural beauty of the region.

After studying this unit you should be able to.

1. Explain the factors favoring karst formation.
2. Identify the locational pattern of karst regions.
3. Describe the landscapes found in karst regions.
4. Understand the subsurface processes which give rise to cave formation.

(B) KEY TERMS AND CONCEPTS

columns	stalactites
karst	stalagmites
sinkholes	uvala

(C) MULTIPLE CHOICE

_____ 1. Beyond limestone, what other rock type can contain karst landforms? (p. 453)

(a) gneiss
(b) shale
(c) dolomite

_____ 2. How does vegetation contribute to the creation of karst formation? (p. 453)

(a) it releases oxygen
(b) it releases carbon dioxide
(c) its roots weaken the rock structure

_____ 3. What type of sinkhole is described when an overlying layer of unconsolidated material is left unsupported and it collapses? (p. 456)

(a) collapse sinkhole
(b) solution sinkhole
(c) suffosion sinkhole

_____ 4. What do we call the irregular, often steep-sided depressions between karst towers? (p. 457)

(a) border
(b) cockpit
(c) uvala

_____ 5. What mineral is formed when dripping water that contains calcium carbonate precipitates its calcite? (p. 458)

(a) travertine
(b) ketchup
(c) carbonic acid

(D) TRUE OR FALSE

1. Karst always develops best in flat landscapes. (p. 453)
T _____ F _____

2. The greater the permeability of the rock, the more susceptible it is to solution, removal, and karst formation. (p. 453)
T _____ F _____

3. In karst topography surface streams are often interrupted and sometimes even stop in mid-valley. (p. 455)
T _____ F _____

4. Karst areas sometimes contain pockets of groundwater that are situated above the level of the local water table. (p. 455)
T _____ F _____

5. The solution features of temperate karst landscapes, tend to be larger than those in tropical regions. (p. 456)
T _____ F _____

(E) SHORT-ANSWER QUESTIONS

1. What makes cave formation in karst landscape different from other caves? (p. 451)

2. How does the nature of the soil and the vegetation influence karst formation? (p. 453)

3. Why is karst most likely found in warm, moist climates? (p. 453)

4. What type of relief is best for karst development? Explain why? (p. 453)

5. How does uplift affect karst formation? (p. 453)

6. Describe one method which earth scientists use to discover the nature of the cave systems. What has this type of research indicated? (p. 455)

7. What does the presence of a **swallow hole** indicate? (p. 456)

8. Distinguish between a **collapse sinkhole** and a **suffosion sinkhole.** (p. 456)

____

____

9. In what climate zones do **uvalas** occur? Why there? (p. 456)

____

____

10. How does the amount of carbon dioxide influence the landforms in karst regions? (p. 458)

____

____

(F) MATCHING

Match the following terms and their meanings:

_______	1. uvala	a.	a deposited pillar on the floor of caves
_______	2. perched aquifers	b.	the joining of two adjacent sinkholes to form a larger depression
_______	3. porosity	c.	irregular, steep-sided depressions between karst towers
_______	4. stalagtites	d.	calcium carbonate precipitating calcite
_______	5. cockpit	e.	icicle-like spikes hanging from cave ceilings
_______	6. stalagmites	f.	water-holding capacity of rock
_______	7. travertine	g.	pockets of water situated above the level of the local water table

(G) ESSAY QUESTIONS

1. Describe the three types of water movement necessary for karst formation.

(i)

(ii)

(iii)

2. Explain how the groundwater in karst landscapes differs from that in non-karst areas.

(H) GRAPHICACY

1. Refer to the topographic map in Figure 44-4.
   a. How is the relief of a map determined?

   b. What is the relief on this map?

c. Draw the map symbol for a depression.

d. Describe the northern half of the map.

2. On the following map of the Adriatic Sea region, draw in the countries of Bosnia, Slovenia, and Croatia and the area where karst landscapes are found.

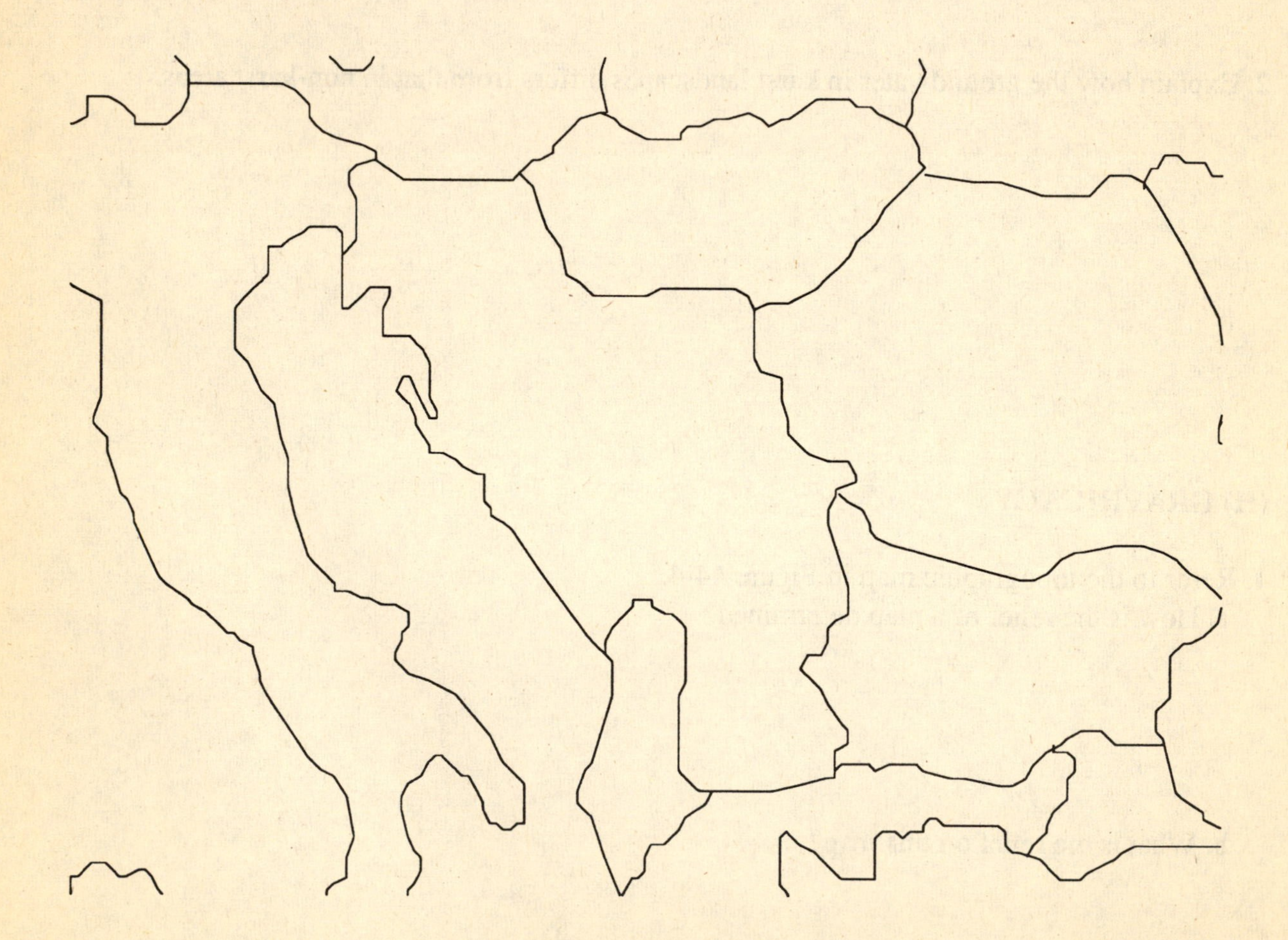

(I) HYPOTHESIS CONSTRUCTION

1. Study the photograph in Figure 44-1 and answer the questions based on what you can see. in the photograph.

a. What negative impact do karst existence have on economic activity?

b. What positive impact could karst have on a region's economy?

c. Write a short (6 line) Chamber of Commerce brochure description that could be used to attract people and economic development to the place photographed in Figure 44-1.

Unit 45

Glacial Degradation and Aggradation

(A) SUMMARY AND OBJECTIVES

Large areas of the earth's surface have only recently emerged from an extensive ice cover that left in its wake, a series of new landforms and landscapes. Glaciation, as the enlargement of the extent of ice cover during periods of earth cooling, has left us with dramatically altered landscapes and drainage patterns. Through an examination of the scientific record, this unit presents a chronology of earth glaciation and a listing of possible explanations for the existence of earth's numerous ice ages. The reasons for the existence of glacial ages are not precisely clear as they range from geomorphological events such as increased volcanic activity to our relationship to the sun. While no one explanation is accepted by all scientists, what is certain is that we are now in a warmer, interglacial period where the human species has itself become a major agent of change.
This warming trend has allowed humans to extend their territory and to develop the divergent patterns of social organization. We are in many ways still creatures of our landscapes.

After studying this unit you should be able to:

1. Understand how glaciers are formed.
2. Discuss and evaluate varied reasons for glaciation.
3. Explain the nature of glacial movement.
4. Describe the erosional capabilities of glaciers.

(B) KEY TERMS AND CONCEPTS

abrasion
basal ice
crevasses
firn
glaciation
interglaciation
roches moutonnee
zone of ablation

(C) MULTIPLE CHOICE

_____ 1. What do we call our present epoch? (p. 461)
(a) Eocene
(b) Pliocene
(c) Holocene

_____ 2. What is the name of our present ice age? (p. 462)

(a) Pleistocene
(b) Cenozoic
(c) Paleocene

_____ 3. What is the name for compacted snow and ice that forms the basis for glaciers? (p. 462)

(a) firn
(b) slurpee
(c) cirque

_____ 4. What term describes the scratches that boulders and rocks make on glaciers? (p. 466)

(a) sizzle marks
(b) striations
(c) benchmarks

_____ 5. What is the hardness level of ice, on the Mohs scale that we covered in Unit 30? (p. 466)

(a) 8
(b) 5.5
(c) 1.5

(D) TRUE OR FALSE

1. When the glaciers expand, the flooded continental shelves will again be exposed. (p. 462)
T _____ F _____

2. Snow pack in polar areas needs to thicker than in temperate areas in order to be converted into glacial ice. (p. 462)
T _____ F _____

3. The compression and recrystallization of snow takes longer in cold polar areas than in moister temperate regions. (p. 462)
T _____ F _____

4. The zone of ablation is the area where the compaction of snow and ice occurs. (p. 463)
T _____ F _____

5. When the lower glacial ice is at the melting temperature, then the glacier moves faster. (p. 465)
T _____ F_____

(E) SHORT-ANSWER QUESTIONS

1. Distinguish between **deglaciation** and **interglaciation.** (p.461)

2. How do the authors describe the present period of earth climate? (p. 461)

3. What makes the present period distinctive from earlier ones? (p. 462)

4. Explain what is presently occurring with the ocean level. (p. 462)

5. Why does glaciation require more time in cold polar areas than in moister temperate zones? (p. 465)

6. Why does the zone of ablation change its position? (p. 464)

7. How does the temperature of basal ice influence the movement of the glacier? (p. 465)

8. Compare the erosional capabilities of temperate zone glaciers with those in polar areas. (p. 465)

____

____

9. Compare glacial creep and glacial sliding. (p. 465)

____

____

10. What is **abrasion** and why is it so important for glacial erosion? (p. 466)

____

____

(F) MATCHING

Match the following terms and their meanings:

_______	1. ablation	a.	large cracks in the glacial ice
_______	2. basal ice	b.	granular, compacted snow and ice
_______	3. glaciation debris	c.	scratches in the bedrock made by a passing glacier
_______	4. crevasses	d.	fragments of bedrock pulled from the surface by the glacier
_______	5. firn	e.	fragments of bedrock are pulled from the surface as the ice moves
_______	6. striations	f.	loss of glacier material through melting and evaporation
_______	7. quarrying	g.	the lowest ice layer

(G) ESSAY QUESTIONS

1. Explain how glaciation influences the planet's radiation and heat balance.

2. Describe the process whereby snow becomes a glacier.

3. Use the 1992 New York Times index in your library and determine what impact the 1991 eruption of the Mt. Pinatubo volcano in the Philippines has had on global climate.

(H) GRAPHICACY

1. Refer to Figure 45-8.

a. Describe the movement of the roche moutonnees as illustrated in the sketches.

b. Why does the roche moutonnees pluck rock on the leeside?

(I) HYPOTHESIS CONSTRUCTION

1. List five reasons that have been given to explain the existence of ice ages.

2. Write out a hypothesis explaining the ice age; your statement is to include at least four of these explanations as independent variables. Your statement should follow the approach given earlier:

$$DV = IV + IV + IV + IV + IV$$

Unit 46

Landforms and Landscapes of Continental Glaciers

(A) SUMMARY AND OBJECTIVES

When author Jean Auel writes about mammoth hunters at the 'dawn of humanity', she is describing a time when parts of the earth were covered by giant glaciers that creaked and groaned as they moved across the surface. Auel's tales vividly capture a time when glaciers did cover large parts of Europe and North America. Today, with warmer temperatures, those glaciers have retreated towards the poles, but we are still left with glacial reminders on our landscape and on our seas where icebergs sometimes lurk. The principles of glacial action that were introduced in the previous unit are now expanded with a focus on continental glaciation. Antarctica and Greenland are not merely far-off and exotic places, but should be seen as regions that play an important role in maintaining global heat balance and sealevel, both of which could change with pronounced global warming. Attention is also paid here to a glacially created example that is much closer to human populations - the Great Lakes region which owes its present formation to glacial scouring and deposition in the last 60,000 years. An additional theme in this unit is that the surface features we observe in the landscape often reveal a great deal about the earth's past as well as hint at possible future changes.

After studying this unit you should be able to:

1. Contrast glaciation at the two polar areas.
2. Differentiate the distinct periods of glaciation.
3. Explain how the Great Lakes were formed.
4. Describe the landforms created by continental glaciers.

(B) KEY TERMS AND CONCEPTS

calving	Laurentide Icesheet
drumlins	moraine
eskers	outwash plain
ice shelves	till

(C) MULTIPLE CHOICE

_____ 1. In size, how large is Greenland when compared to Antarctica? (p. 473)
(a) about one-half as large
(b) about one-quarter as large
(c) about one-eighth as large

_____ 2. What is the average length of time between glaciations? (p. 474)
(a) 65,000 years
(b) 95,000 years
(c) 250,000 years

_____ 3. What is the name of the largest continental glacier in Eurasia? (p. 474)
(a) European
(b) Russian
(c) Scandinavian

_____ 4. Which was the most recent of the North American glacial advances? (p. 475)
(a) Wisconsin
(b) Kansan
(c) Nebraskan

_____ 5. Which glacial landform has a flat-top and can be compared to a delta? (p. 481)
(a) kettle
(b) kame
(c) esker

(D) TRUE OR FALSE

1. The Antarctic Icesheet moves outward in a simple radial pattern. (p. 471)
T _____ F _____

2. Calving is another form of ablation. (p. 471)
T _____ F _____

3. Pack ice is ice that has broken off from the icesheet. (p. 471)
T _____ F _____

4. Till is the unsorted mass of material carried and deposited by glaciers. (p. 478)
T _____ F _____

5. Drumlins are glacially-created hills that lie perpendicular to the glacier. (p. 479)
T _____ F _____

(E) SHORT-ANSWER QUESTIONS

1. Why are the polar areas regarded as deserts? (p. 470)

____

2. Explain what would happen to the Antarctic itself if the icesheet were to melt? (p. 470)

____

____

3. What are **flow regimes**? (p. 471)

____

4. How does **pack ice** differ from icesheets? (p. 471)

____

____

5. What icesheet data do scientists obtain from deep-sea sediments? (p. 473)

____

6. What are varves and what do they tell us about glaciation? (p. 476)

____

____

7. What does the existence of **pluvial lakes** tell us about glaciation in an area? (p. 476)

____

____

8. How does **recessional moraine** differ from terminal moraine? (p. 479)

____

9. What can be deduced about glacial movement when we study a **drumlin**? (p. 479)

____

10. How does the **outwash plain** differ from the area under the glacier? (p. 480)

____

____

(F) MATCHING

Match the following terms and their meanings:

_______	1. calving	a.	long sinuous ridge left by glacial melting
_______	2. drumlins	b.	material transported far from their source by glaciers
_______	3. nunataks	c.	paired layers of lake-bottom sediments
_______	4. varves	d.	the breaking up of iceshelves into icebergs
_______	5. eskers	e.	a small regional ice mass
_______	6. pack ice	f.	smooth, elliptical hill composed of glacial drift
_______	7. icecap	g.	mountain peaks protruding through continental glaciers
_______	8. erratics	h.	floating sea ice

(G) ESSAY QUESTIONS

1. Describe the changes in scientific methods used to determine the succession of glaciations.

2. What does study of the Great Salt Lake in Utah tell us about glaciation?

3. Describe how the Great Lakes were formed.

(H) GRAPHICACY

1. Compare the scales of the following maps - Antarctica in Figure 46-1 and North America on page 291 of the text.

a. Note the differences in scale between the two maps.

b. What is the relationship in size between Antarctica and the lower 48 U.S. states?

c. How does the Ross Ice Shelf compare in size to the lower 48 U.S. states?

d. What are the climatic and human implications if the Ross Ice Shelf breaks off and melts?

(I) HYPOTHESIS CONSTRUCTION

1. Alfred Wegener of continental drift fame did meteorological research on the Greenland Icecap. He died there during a snowstorm while on one of his expeditions. Why would one choose this area for weather and climate research?

2. Imagine that you have been chosen to do funded research at a scientific station in Greenland. Explain the type of research that you would wish to do.

Unit 47

Landforms and Landscapes of Mountain Glaciers

(A) SUMMARY AND OBJECTIVES

In many ways, mountain glaciers are the most powerful agents of landscape change in their respective regions. The force of gravity ensures that the topography is forever changed once a glacier has passed through. The glaciers we examine in this unit are located on most of the continents and are all found above 4,400 m (14,500 ft). Most of the landscapes in these glaciated regions have been created by the powerful glacial erosional force which uses gravity and the mass of compacted snow to carve distinctive, often jagged landforms. The landscapes, though scenic and good for tourism, do limit other economic activity.

After studying this unit you should be able to:

1. Describe the global distribution of mountain glaciers.
2. Explain how mountain glaciers differ from continental glaciers.
3. Understand how distinctive glacial landforms develop.
4. Describe the impact of glacial landscapes on human activity.

(B) KEY TERMS AND CONCEPTS

arete	medial moraine
cirque	moraine
fjord	rock flour
hanging valley	tarn

(C) MULTIPLE CHOICE

_____ 1. Which major landmass does not contain alpine glaciers? (p. 483)
(a) Asia
(b) Australia
(c) Africa

_____ 2. Where is the Beardmore Glacier found? (p. 484)
(a) Alaska
(b) Antarctica
(c) the Swiss Alps

_____ 3. What is the latitudinal limit for most glacier formation? (p. 484)

(a) poleward of 65 degrees
(b) poleward of 30 degrees
(c) poleward of 50 degrees

_____ 4. What do we call the bowl-shaped, steep depressions where an alpine glacier starts? (p. 487)

(a) arete
(b) cirque
(c) tarn

_____ 5. What term describes the area where the alpine glacier melts and evaporates? (p. 489)

(a) ablation zone
(b) accumulation zone
(c) retreat zone

(D) TRUE OR FALSE

1. Moraine lakes are small lakes which form on the floors of old cirques. (p. 488)
T _____ F _____

2. All of Africa's glaciers are found on either Mt. Kilimanjaro or on Mt. Kenya. (p. 484)
T _____ F _____

3. The early Cenozoic era was one of the coldest in earth history. (p. 482)
T _____ F _____

4. The valley of a glacier is very similar in shape to a river valley in the floodplain stage. (p. 486)
T _____ F _____

5. Valley train is the term which describes the alluvium derived from meltwater containing morainal material. (p. 490)
T _____ F _____

(E) SHORT-ANSWER QUESTIONS

1. What are the lower altitudinal limits for glaciers within the tropics? (p. 484)

____

2. Give three reasons why New Zealand has glaciers while nearby Australia does not. (p. 484)

____

____

3. Compare the shape of a glacial valley with a stream valley. (p. 486)

____

____

4. How does a **truncated** spur get created? (p. 486)

____

____

5. Why are the valley floors of main glaciers and the floors of their tributary glaciers at different elevations? (p. 487)

____

____

6. Describe the form and origin of **finger lakes**. (p. 488)

____

____

7. What do **rock steps** reveal about the glacial process? (p. 488)

____

____

8. How are **fjords** formed? (p. 488)

____

____

9. What is **rock flour** and what happens to it eventually? (p. 489)

____

____

10. Distinguish between lateral and medial moraine. (p. 489)

____

____

(F) MATCHING

Match the following terms and their meanings:

_______	1. cirque	a.	a jagged ridge separating adjacent glacial valleys
_______	2. horn	b.	a narrow steep-sided estuary formed from a glacial trough
_______	3. arete	c.	a lake formed in a glacial cirque
_______	4. moraine	d.	very fine particles of glacial debris
_______	5. fjord	e.	a steep-sided sharp-edged glacial peak
_______	6. rock flour	f.	debris left by the glacier
_______	7. tarn	g.	an amphitheater-shaped landform caused by glacial erosion

(G) ESSAY QUESTIONS

1. How does the erosional force of a mountain glacier differ from that of a continental glacier?

(H) GRAPHICACY

1. On the following contour sketch map, identify the features.
   a. glacial trough
   b. truncated spur
   c. cirque

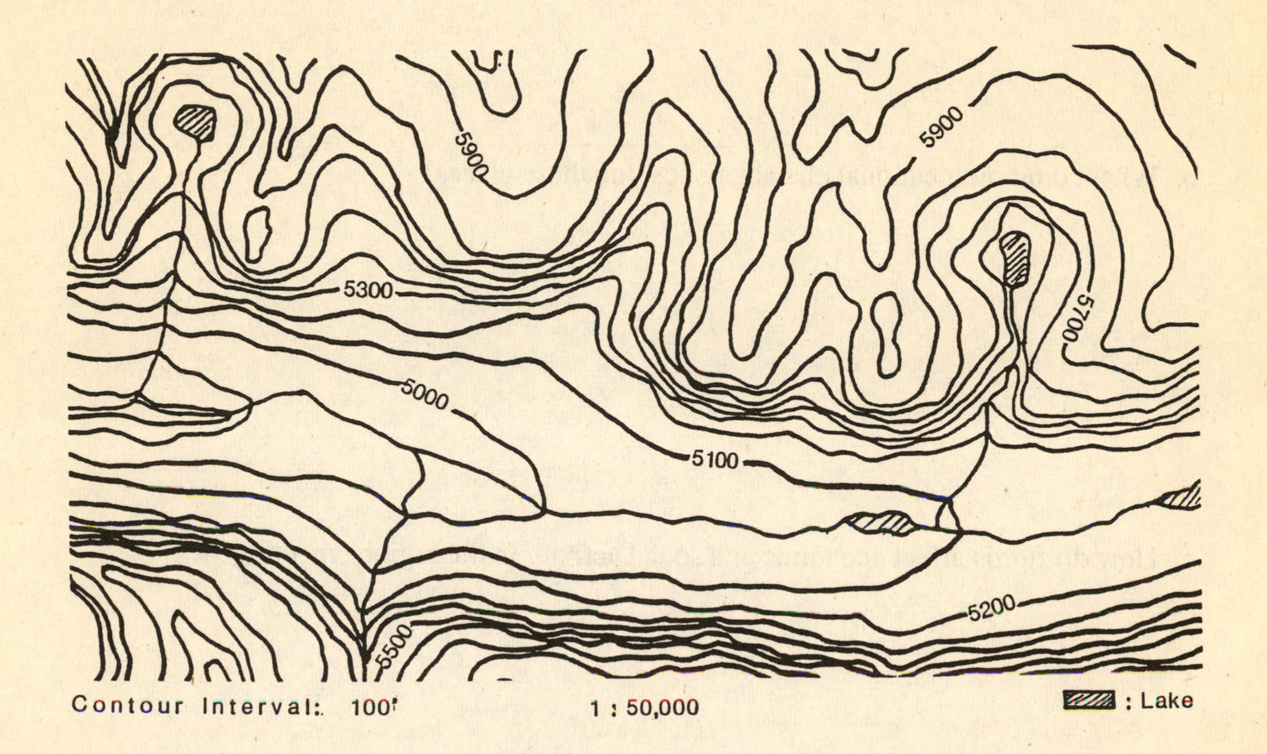

2. Refer to the opening photo on page 482 in the text.

   a. Identify the following features:
      (i) medial moraine
      (ii) lateral moraine
      (iii) truncated spur
      (iv) hanging valley
      (v) zone of accumulation

   b. In what directions would this glacier *not* be flowing? Explain.

(I) HYPOTHESIS CONSTRUCTION

1. Fjords are a significant part of the landscape in many glacial areas.

   a. Name three regions of the world where fjords be found.

   b. What common locational characteristics link these places?

   c. How do fjords affect economic and social activity in the regions mentioned?

Periglacial Environments and Landscapes

(A) SUMMARY AND OBJECTIVES

While climate always has an impact on the landforms of any region, this influence is perhaps most pronounced in high latitude areas that experience some of the coldest, most severe temperature conditions. This unit examines those periglacial regions where the subsurface ground is permanently frozen and where weathering and mass movement processes create a very distinctive landscape. Such ecologically fragile areas are thinly inhabited but are now in danger of disruption with the greater human search for energy sources. We are all challenged in our role as earth stewards to better understand the fragility of these environments and to be wary of the dangers that human intrusion such as mineral exploration and oil spills will create. We are again forced to consider the balance between the needs of an industrial era and the preservation of the earth's wilderness areas.

After studying this unit you should be able to:

1. Explain how periglacial regions are distinctive.
2. Describe the many unique periglacial landforms.
3. Discuss the weathering and mass movement processes in periglacial areas.
4. Understand the impact that human activity can have in periglacial environments.

(B) KEY TERMS AND CONCEPTS

frost heaving	permafrost
frost wedging	pingo
periglacial	solifluction

(C) MULTIPLE CHOICE

_____ 1. Which of these regions have no periglacial environments? (p. 491)

(a) Antarctica
(b) northern Europe
(c) southern Africa

_____ 2. Which of the following climate types is *not* associated with a periglacial environment? (p. 491)

(a) Dfd
(b) Dw
(c) E

_____ 3. What percent of the world's land area is dominated by periglacial conditions? (p. 491)

(a) one-half
(b) one-quarter
(c) one-third

_____ 4. In which vegetation zone is continuous permafrost predominant? (p. 493)

(a) tundra
(b) forest
(c) taiga

_____ 5. Which type of frost action moves rock fragments horizontally within the active layer? (p. 494)

(a) frost wedging
(b) frost thrusting
(c) frost creep

(D) TRUE OR FALSE

1. The upper surface of the permafrost is called the permafrost table. (p. 492)
T _____ F _____

2. Frost wedging is greatest when the rock contains very little water. (p. 493)
T _____ F _____

3. Frost heaving involves the movement of particles in the active layer under the influence of gravity. (p. 494)
T _____ F _____

4. Pingo is an Inuit word for a hill. (p. 496)
T _____ F _____

5. Felsenmeer is another name for a boulder field. (p. 497)
T _____ F_____

(E) SHORT-ANSWER QUESTIONS

1. How do **periglacial** areas differ from glacial ones? (p. 492)

2. How does **permafrost** develop in periglacial regions? (p. 492)

3. Distinguish between the permafrost table and the active layer. (p. 492)

4. How does the thickness of the **active layer** change with latitude? (p. 492)

5. What prevents the permafrost from constantly thickening? (p. 493)

6. What is the difference between continuous and discontinuous permafrost? (p. 493)

7. How does rock porosity influence frost wedging? (p. 493)

8. How does the texture of the soil influence solifluction? (p. 495)

____

____

9. Why do periglacial landscapes often have distinctive polygon patterns on the ground? (p. 495)

____

____

10. Describe **patterned ground** in periglacial environments. (p. 496)

____

(F) MATCHING

Match the following terms and their meanings:

_______	1. solifluction	a.	rounded hill with a core of ice
_______	2. permafrost	b.	the movement of particles under the influence of gravity
_______	3. pingo	c.	slopes covered by blocky pieces of rock
_______	4. frost creep	d.	permanently frozen ground
_______	5. boulder fields	e.	a slow flowage of saturated soil above the permafrost layer

(G) ESSAY QUESTIONS

1. Describe how landscape modification in periglacial areas differs from that in other zones.

2. How does human activity influence periglacial ecology?

(H) GRAPHICACY

1. On the world map, draw in the periglacial regions.

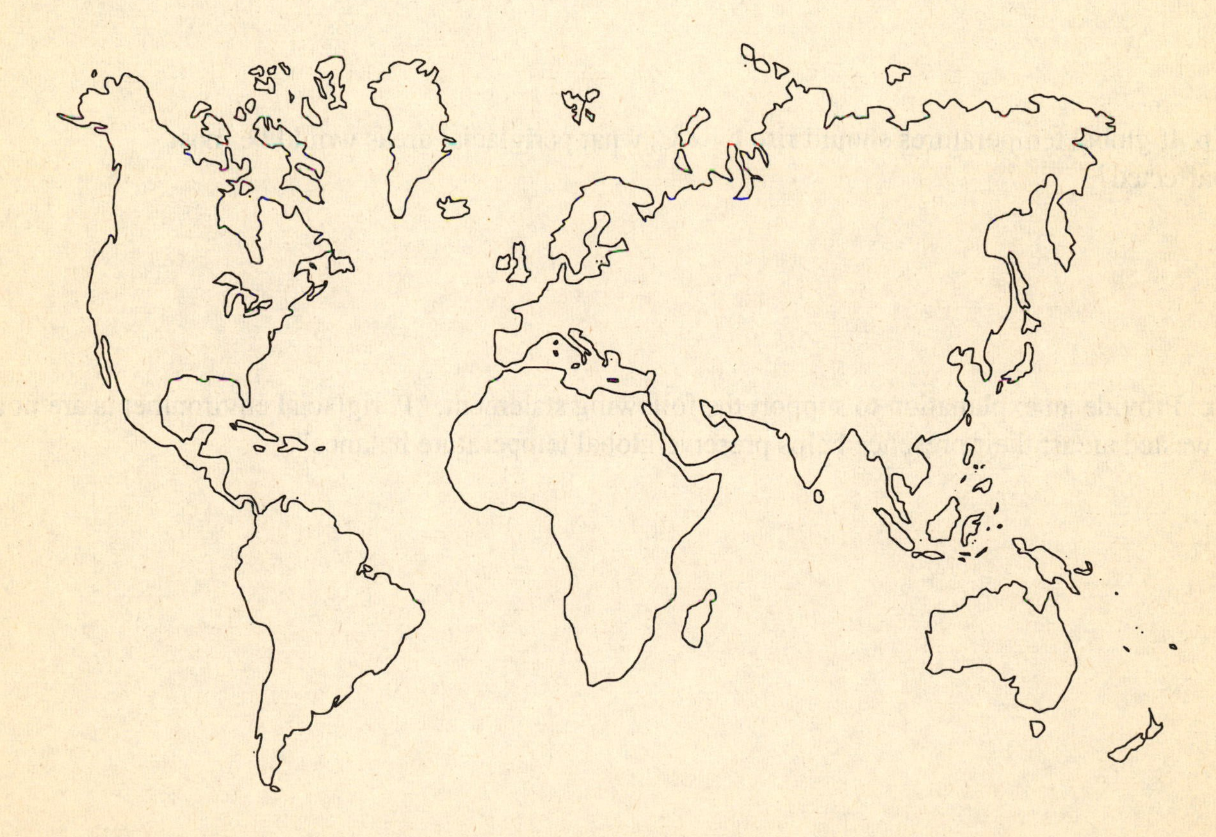

a. There should be some areas on your map that are not continuous. Explain why these are periglacial.

b. Draw in the Alaskan Pipeline. Why was the pipeline constructed above ground?

(I) HYPOTHESIS CONSTRUCTION

1. Compare the map of periglacial regions constructed in part H with the global temperature maps on page 93 of the text.

a. What are the present seasonal temperature limits for periglacial conditions?

b. If global temperatures should rise by 2^{0}C, what periglacial areas would be most affected?

c. Provide an explanation to support the following statement: "Periglacial environments are not wasted areas; their presence helps preserve global temperature balance".

Unit 49

Wind as a Geomorphic Agent

(A) SUMMARY AND OBJECTIVES

Wind is such a constant in our environment that we often overlook its ability to create new landforms and to change existing landscapes. As an aggradational force, wind can reform dry earth surfaces with high velocity and through a constant direction, create distinctive dunes. At the same time, wind also has the ability to transport material over great distances and to deposit it in new landscapes. The most significant wind-borne material is a fine sediment of glacial origin called loess. This productive mineral mixture is found in agricultural areas stretching from the central United States to the plains of North China. In all areas loess is particularly fertile and is used extensively in the world's breadbaskets.

After studying this unit you should be able to:

1. Understand the nature of wind erosion and deposition.
2. Describe the landforms created by wind.
3. Explain loess material and its global distribution.
4. Appreciate the economic significance of loess.

(B) KEY TERMS AND CONCEPTS

abrasion	erg
barchan	loess
dune	saltation
eolian	yardang

(C) MULTIPLE CHOICE

_____ 1. What is the name given to a sand sea? (p. 502)
(a) reg
(b) erg
(c) claude

_____ 2. What is the leeward side of a dune called? (p. 503)
(a) slip face
(b) glide side
(c) backslope

_____ 3. Which side of the barchan is convex shaped? (p. 503)

(a) both
(b) leeward
(c) windward

_____ 4. Which of these dunes are at right angles to the wind direction? (p. 504)

(a) longitudinal
(b) parabolic
(c) transverse

_____ 5. Which of these states has the greatest concentration of loess soil? (p. 505)

(a) Arizona
(b) Georgia
(c) Nebraska

(D) TRUE OR FALSE

1. Yardangs are perpendicular to the prevailing wind direction. (p. 501)
T _____ F _____

2. The windward side of the dune is named the backslope. (p. 503)
T _____ F _____

3. In the parabolic dune the windward side has a concave form. (p. 504)
T _____ F _____

4. In loess soil the fertility is only found in the top layer. (p. 506)
T _____ F _____

5. Loess soil has few well developed layers and it is highly porous. (p. 506)
T _____ F _____

(E) SHORT-ANSWER QUESTIONS

1. What are the two wind variables that determine wind erosion? (p. 499)

____

____

2. Describe the process of **wind abrasion**. (p. 501)

____

____

3. How are **deflation hollows** formed? (p. 501)

____

____

4. Under what conditions do **yardangs** develop? (p. 501)

____

____

5. Distinguish between suspension and saltation. (p. 501)

____

____

6. In which biomes are wind-developed landscapes most pronounced? (p. 502)

____

7. Distinguish between active and fixed dunes. (p. 503)

____

____

8. What makes a **barchan** so distinctive? (p. 503)

____

____

9. Name two regions outside the U.S. where loess is found extensively. (p. 505)

____

10. What does loess contain that makes it so valuable? (p. 506)

____

(F) MATCHING

Match the following terms and their meanings:

_______ 1. yardang		a. a sand sea
_______ 2. eolian erosion		b. a surface of closely packed pebbles
_______ 3. erg		c. low sand ridges parallel to the prevailing wind direction
_______ 4. deflation		d. movement of small particles in wind by a series of jumps
_______ 5. desert pavement		e. a crescent-shaped dune
_______ 6. saltation		f. wind-related process
_______ 7. barchan		g. movement of loose, fine particles

(G) ESSAY QUESTIONS

1. Describe how glacial landscapes are affected by wind modification.

2. Compare the characteristics of parabolic and longitudinal dunes, including their shape, wind requirements, and distribution.

(H) GRAPHICACY

1. Refer to the unit opening photo on page 499 of the text.

a. What type of dunes are these?

b. Identify and name three features of these dunes.

c. What is the dominant wind direction in this scene?

d. This is a scene in Africa's Namib Desert. Name three other places where similar scenes could be photographed.

(I) HYPOTHESIS CONSTRUCTION

1. While loess is a fertile basis for agriculture, its value depends on other variables such as precipitation and evaporation rates. Compare the map of loess distribution in Figure 49-9 of the text to the map of precipitation in Figure 12-10 on page 135.

a. List the regions with major loess deposits.

b. List the four leading agricultural regions which are loess-based.

c. Describe the precipitation limits for these agricultural areas.

d. Construct a hypothesis relating the presence of loess, agricultural potential, and precipitation. Your dependent variable should be agricultural productivity.

Unit 50

Coastal Processes

(A) SUMMARY AND OBJECTIVES

Shorelines and coastal areas are a major area of research for physical geographers because they represent an important zone of interaction between physical forces and major human settlement patterns. Coastlines can be seen as products of the energy contained in the oceans. This energy finds its expression as wave action which both erodes and deposits material along our coastlines. While most attention is on the relentless role of waves, the power is seen to be mediated by the pattern of tides which periodically vary the coastal water level. In addition, storm surges and crustal movement are presented as significant forces which increase the ferocity of wave action and produce new landforms in a shorter timeframe. Our interest in coastlines is largely due to the increasing human population that resides at or near the coast.

After studying this unit you should be able to:

1. Identify the importance of coastal zones for the study of physical geography.
2. Describe the properties of waves.
3. Explain the causes and impacts of tides.
4. Understand the role of storms in increasing coastal erosion.

(B) KEY TERMS AND CONCEPTS

barrier islands	shore
corrasion	storm surge
longshore drift	swash
shoaling	wave refraction

(C) MULTIPLE CHOICE

_____ 1. What is the term used to describe the narrow belt of land bordering a body of water? (p. 509)

(a) coast
(b) shore
(c) shoreline

_____ 2. What do we call the distance over which ocean winds blow? (p. 509)

(a) period
(b) grout
(c) fetch

_____ 3. What term describes the impact of shallow water on an advancing wave? (p. 511)

(a) shoaling
(b) breaking
(c) curling

_____ 4. Which type of tide has the least extreme tides? (p. 514)

(a) neap
(b) spring
(c) low

_____ 5. Which type of current flows from the shore seaward? (p. 515)

(a) rip
(b) longshore
(c) refracted

(D) TRUE OR FALSE

1. The wave length is the vertical distance between the top of a wave and its bottom. (p. 509)
T _____ F _____

2. Swash is when waves lose their form and the water slides up the beach. (p. 511)
T _____ F _____

3. When the earth, moon, and sun are aligned in a straight line, the result is a neap tide. (p. 514)
T _____ F_____

4. During wave refraction when the wave is in shallower water, its angle to the shoreline is much greater than during the deep water period. (p. 511)
T _____ F _____

5. The mechanical erosion process of waves is called corrasion. (p. 513)
T _____ F _____

(E) SHORT-ANSWER QUESTIONS

1. Why is the **littoral zone** significant in the study of physical geography? (p. 509)

____

____

2. What four factors must be present for waves to form? (p. 509)

____

____

3. What is the mathematical relationship between the depth of a wave of oscillation and its length? (p. 509)

____

4. How does coastline shape influence **wave refraction**? (p. 511)

____

____

5. What is the ultimate impact of wave refraction on a coastline? (p. 511)

____

____

6. How does **longshore drifting** change the coastline? (p. 512)

____

____

7. How are **barrier islands** generated? (p. 513)

____

____

8. What are the three forces that control the earth's tides? (p. 514)

____

9. Why are geographers so interested in the phenomenon of tidal range? (p. 515)

____

____

10. How do **rip currents** compare to other shore currents? (p. 515)

____

(F) MATCHING

Match the following terms and their meanings:

_______	1. fetch	a.	a combination of rising water and forceful wave action
_______	2. wave trough	b.	the narrow belt of land bordering a body of water
_______	3. neap tide	c.	mechanical erosional process
_______	4. swell	d.	an unusually high tide
_______	5. shore	e.	the bottom of the wave
_______	6. storm surge	f.	the distance over open water which the wind blows
_______	7. corrasion	g.	long rolling waves traveling great distances

(G) ESSAY QUESTIONS

1. Describe the nature of waves of oscillation.

2. Describe, with the aid of a sketch the concept of wave refraction.

3. Name and describe the three main erosional processes involved in wave action.

(i) ______________

(ii) ______________

(iii) ______________

(H) GRAPHICACY

1. What are barrier islands?

2. Examine a map of the United States and identify three major barrier islands.

3. What problems could await these barrier islands should climate become warmer?

(I) HYPOTHESIS CONSTRUCTION

1. Write a one-sentence statement explaining why coastal processes will become an increasingly important topic for geographers.

2. Explain how humans try to alter the effects of beach drifting.

3. What are the positive and negative long-run implications?

Unit 51

Coastal Landforms and Landscapes

(A) SUMMARY AND OBJECTIVES

After examining the nature of wave action in the previous unit, here we identify and define the various landforms that make up coastal landscapes. The processes of wave erosion and deposition constantly form and reform coasts giving us features such as dunes, sea arches, barrier islands, and sandspits. Other forces include tectonic plate movement which changes the land/sea spatial relationship and limestone deposition which gives us tropical coral reefs. Significant human/land interaction occurs along coasts and the choices for settlement patterns often neglect the longer term trends of changes in the physical world. Global climate change and its increase of temperature will invariably lead to increased sea levels and greater stress on communities settled at the waters edge.

After studying this unit you should be able to:

1. Understand the changing nature of the beach.
2. Describe the formation of landforms along beach areas.
3. Explain how tectonic plate movements affect beach landscapes.
4. Identify the dangers of barrier beach development.

(B) KEY TERMS AND CONCEPTS

atoll
barrier island
berm
coral reef
sandspit
stack
tombolo
wave-cut platform

(C) MULTIPLE CHOICE

_____ 1. What mineral gives beach sand its light color? (p. 519)
(a) coral
(b) limestone
(c) quartz

_____ 2. What term is used for sandy beaches that were laid out during storms and are beyond the reach of normal wave action? (p. 519)

(a) backshore
(b) bars
(c) berm

_____ 3. What are sandspits called when they grow all the way across the mouth of a bay? (p. 520)

(a) baymouth bars
(b) blocking bars
(c) juice bars

_____ 4. What term describes the nearly flat bedrock eroded surface at the base of cliffs? (p. 524)

(a) cut-back platform
(b) wave-cut platform
(c) berms

_____ 5. When did the previous rise in ocean level start slowing down? (p. 526)

(a) 200 years ago
(b) 5,000 years ago
(c) 15,000 years ago

(D) TRUE OR FALSE

1. The summer waves produce a narrower berm than at other times of the year. (p. 519)
T _____ F _____

2. Some barrier islands originated as offshore bars during the last glaciation. (p. 522)
T _____ F _____

3. The coast of Oregon consists mostly of hard crystalline rocks and they wear away much slower than other coasts. (p. 524)
T _____ F _____

4. More coasts are of an emergent than a submergent origin. (p. 525)
T _____ F _____

5. Charles Darwin concluded that the coral reefs had developed on the rims of eroded volcanic cones. (p. 526)
T _____ F _____

(E) SHORT-ANSWER QUESTIONS

1. Why is the beach normally wider than what we initially see? (p. 519)

____

____

2. Why do the U.S. East Coast and Hawaii have different colored beach sand? (p. 519)

____

3. Why does the U.S. Northwest Coast have little sand? (p. 519)

____

4. Describe the **foreshore** area. (p. 519)

____

____

5. How do beach profiles differ from summer to winter? (p. 519)

____

____

6. How do **offshore bars** affect coastal erosion? (p. 521)

____

____

7. Why do lagoons behind barrier islands rarely become swampy? (p. 522)

____

8. What processes are indicated by the presence of sea arches and sea stacks? (p. 524)

____

____

9. Name two examples of coasts of emergence. (p. 525)

____

10. How are coasts of **submergence** created? (p. 525)

____

____

(F) MATCHING

Match the following terms and their meanings:

_______	1. berm	a.	gently sloping surface produced by wave erosion at the base of a cliff
_______	2. tombolo	b.	a ridge of sand parallel to the beach
_______	3. shingle beach	c.	circular coral reefs surrounding a lagoon
_______	4. longshore bar	d.	a sandy backshore beach laid down during storms
_______	5. coral	e.	a gravel and pebble beach
_______	6. atolls	f.	depositional landform connecting an off-shore island with the coast
_______	7. wave-cut	g.	the calcium carbonate remains of marine platform organisms

(G) ESSAY QUESTIONS

1. Describe the changes in coastal ecology when a baymouth forms.

2. Explain the conditions which are necessary for coral reefs to develop.

(H) GRAPHICACY

1. Refer to the opening photo on page 518 of the text.

   a. List four coastal processes that are illustrated in this photograph.

   b. Is this a coast of emergence or submergence? Explain.

2. The following is a sketch of a U.S. barrier island region.

   a. Identify the region.
   b. Indicate the dominant direction of the current.
   c. Indicate the worst site for a settlement.

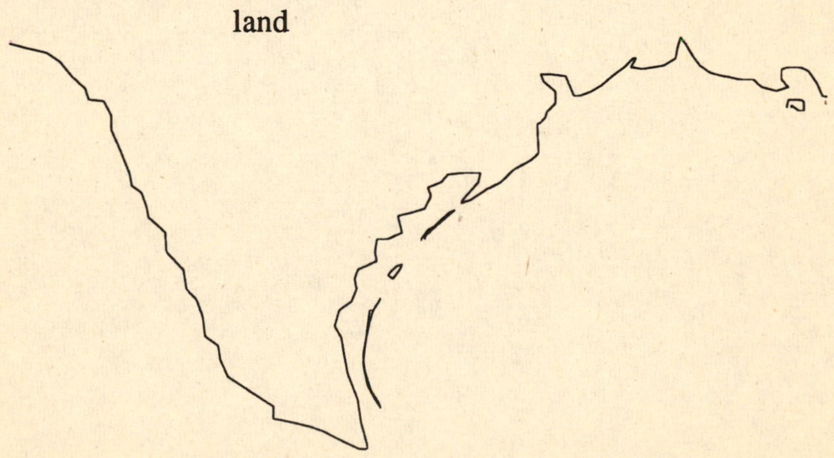

land

(I) HYPOTHESIS CONSTRUCTION

1. Compare the coastlines in Figure 51-5 with that in the unit opening photo on page 518 of the text.

   a. Describe the limitations to human activity in each case.

   b. Describe the benefits each of these coastlines give to human activity.

   c. Complete the following statement. For human beings coastlines are not merely neutral phenomena, they

   ____

   ____

Unit 52

Physiographic Realms: The Spatial Variation of Landscapes

(A) SUMMARY AND OBJECTIVES

As with all scientific fields, physical geography data must be categorized if we are to develop clearer insights into the world around us. Geography's distinctive approach is to develop spatial categories or regions that assist us in explaining phenomena and places we are studying. The physiographic focus in this unit and the next looks at geographic areas in their totality by dividing the world into realms and regions. These physiographic divisions incorporate knowledge of all the sub-fields we have studied thus far including climatology, biogeography, and pedology. Physiography synthesizes all of the elements of the physical realm into regional divisions.

After studying this unit you should be able to:

1. Understand the role of regionalization in geographic study.
2. Differentiate between physiographic realms and regions.
3. Locate the North American realms.
4. Describe the distinguishing features of the North American realms.

(B) KEY TERMS AND CONCEPTS

physiographic realm	piedmont
physiographic region	plains
physiography	shield

(C) MULTIPLE CHOICE

_____ 1. What would be a second-order physiographic unit? (p. 530)
(a) Europe
(b) Scandinavia
(c) northern Europe

_____ 2. How many physiographic realms are found in North America? (p. 532)
(a) 6
(b) 12
(c) 15

_____ 3. What type of rock does the Canadian Shield consist of? (p. 533)
(a) sedimentary and igneous
(b) metamorphic and igneous
(c) metamorphic and sedimentary

_____ 4. Which is North America's most varied physiographic realm? (p. 534)
(a) Appalachian
(b) Interior Plains
(c) Western Mountains

_____ 5. What name is given to the foothills of larger mountain ranges? (p. 536)
(a) piedmont
(b) tarn
(c) lowlands

(D) TRUE OR FALSE

1. The Interior Plains is underlain mostly by sedimentary rocks. (p. 533)
T _____ F _____

2. Histosols form the fertile soils of The Great Plains. (p. 533)
T _____ F _____

3. The Appalachians have a much greater regularity of range direction than the Rockies. (p. 534)
T _____ F _____

4. The Great Smoky Mountains are often referred to as the "Younger" Appalachians. (p. 534)
T _____ F _____

5. Some karst topography is present in central Florida. (p. 536)
T _____ F _____

(E) SHORT-ANSWER QUESTIONS

1. Define **physiography**. (p. 531)

____

____

2. What function does **regionalization** serve for geographers? (p. 528)

____

____

3. How do first order realms differ from the second and third order? (p. 530)

____

4. Name four characteristics of the Canadian Shield. (p. 533)

____

____

5. What are the eastern and western limits of the Interior Plains? (p. 533)

____

6. What glacial feature sustains agriculture on the Interior Plains? (p. 533)

____

7. How does soil in the Great Plains differ from that in the Interior Plains? (p. 533)

____

8. Describe the Ozark Plateau. (p. 534)

____

____

9. What are the general features of the Appalachian Highlands? (p. 534)

____

____

10. Name four features of the Gulf-Atlantic Coastal Plain. (p. 535)

____

____

(F) MATCHING

In which physiographic ***realm*** is each of the following located?

_______	1. Portland, Oregon	a. Appalachian Highlands
_______	2. Yucatan Peninsula	b. Interior Plains
_______	3. Ottawa, Canada	c. Western Mountains
_______	4. Columbus, Ohio	d. Gulf-Atlantic Coastal Plain
_______	5. Gettysburg, Pennsylvania	e. Canadian Shield

(G) ESSAY QUESTIONS

1. Discuss the difficulties earth scientists face when demarcating the boundaries of physiographic realms.

(H) GRAPHICACY

1. "The Great Plains" conjures up many images in your mind. On the following map, draw in the limits of what you perceive to be the Great Plains.

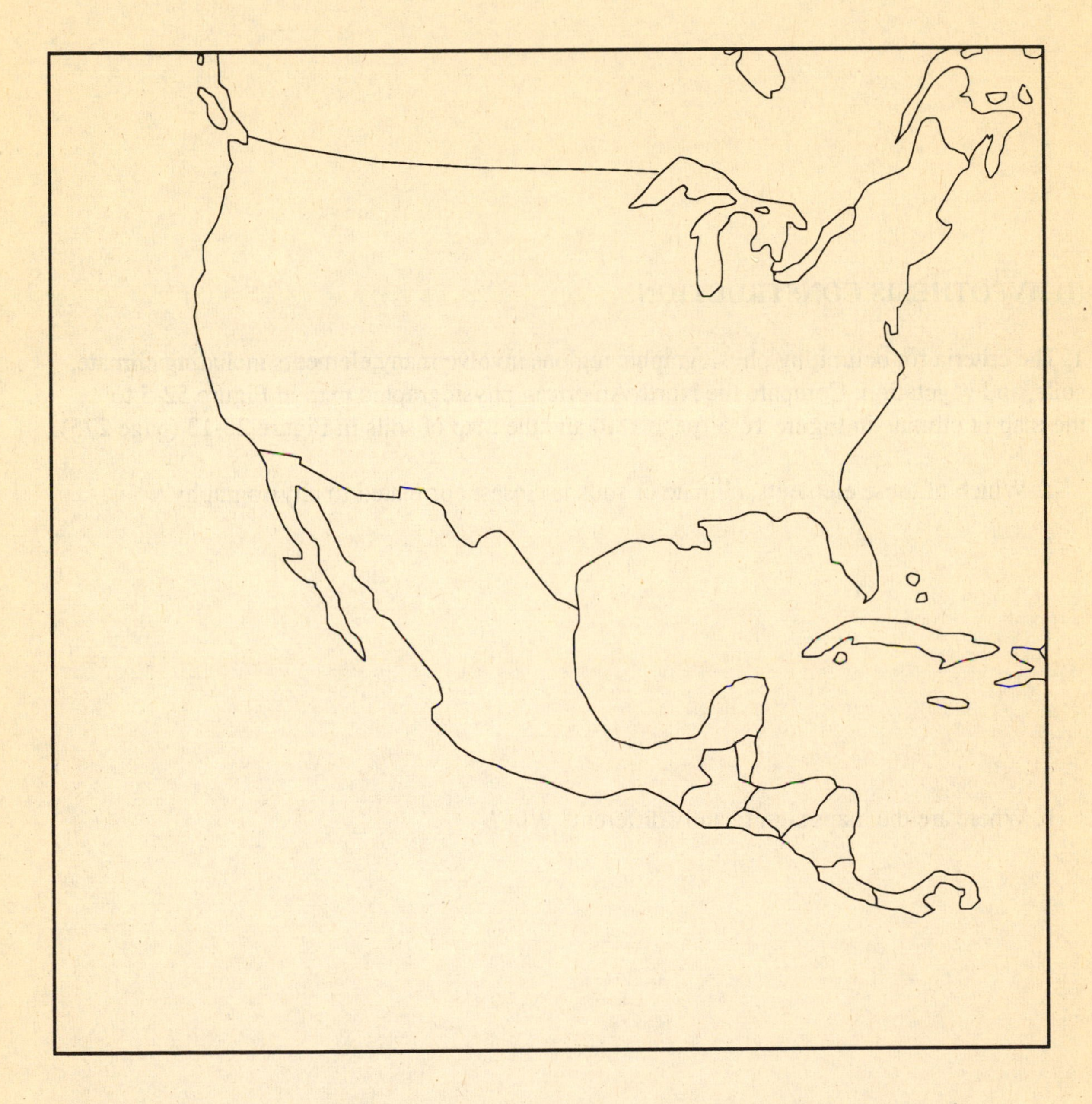

a. Now compare your map with that of someone else in the class. What are the major differences? Why do you have these differences?

b. Discuss the problems faced with regionalization.

(I) HYPOTHESIS CONSTRUCTION

1. The criteria for delimiting physiographic regions involve many elements including climate, soils, and vegetation. Compare the North American physiographic map in Figure 52-5 to the map of climate in Figure 16-3 (page 180) and the map of soils in Figure 25-13 (page 275).

a. Which of these elements, climate or soils is closest correlated to physiography?

b. Where are the maps significantly different? Why?

Unit 53

Physiographic Regions of the United States

(A) SUMMARY AND OBJECTIVES

After the overview of physiographic realms and regions, this unit explores, in greater, detail the distinctive regions within two U.S. realms. The Appalachian Highlands and Western Mountains are analyzed for their internal characteristics and for the role these features have played in settlement patterns. For the geographer, the tie between humans and their environments form a critical foundation of study. We perceive and use the environment in particular ways that vary from culture to culture and from place to place. The study of physical geography allows us to better understand these variations and to care deeper for our global environment.

After studying this unit you should be able to:

1. Locate the major physiographic regions in the United States.
2. Describe the characteristics of the Appalachian Highlands and the Western Mountains.
3. Understand the distinctive regions within these realms.
4. Relate human activity in these physiographic regions to their structural features.

(B) KEY TERMS AND CONCEPTS

Appalachian Plateau
basin-and-range
Blue Ridge Mountains
Colorado Plateau
Columbia Plateau
ridge-and-valley

(C) MULTIPLE CHOICE

_____ 1. In which section of the Appalachians do we find the ridge-and-valley? (p. 539)
(a) northern
(b) central
(c) southern

_____ 2. What is the major type of coal found in the Appalachian Plateau? (p. 539)
(a) lignite
(b) anthracite
(c) bituminous

_____ 3. What is the name of the low passage through the Kentuckian Appalachians that serves as a westward pathway? (p. 540)

(a) Lancaster Gap
(b) Green Grass Gap
(c) Cumberland Gap

_____ 4. What type of rocks are in the Appalachian Great Valley region where Pennsylvania Dutch farmers settled? (p. 541)

(a) dolomite and slate
(b) limestone and slate
(c) slate and dolorite

_____ 5. Which of these regions was strongly affected by glaciation? (p. 541)

(a) New England
(b) Appalachian
(c) Piedmont

(D) TRUE OR FALSE

1. The eastern interior boundary of the Appalachian Highlands is not well defined with escarpments. (p. 540)
T _____ F _____

2. The Blue Ridge Mountains are the older part of the Appalachians. (p. 541)
T _____ F _____

3. The Snake River in Idaho forms the northern limit of the Northern Rocky Mountains. (p. 543)
T _____ F _____

4. The Basin-and-Range Province consists of one large river valley. (p. 544)
T _____ F _____

5. The Pacific Mountains and Valleys lie on an east-west axis. (p. 544)
T _____ F _____

(E) SHORT-ANSWER QUESTIONS

1. How were the Appalachians viewed by early European colonists? (p. 540)

____

____

2. What is the most distinctive feature of the Newer Appalachians. (p. 540)

____

____

3. What rock composition distinguishes the Blue Ridge Mountains from the Appalachian Plateau? (p. 541)

____

4. What is the derivation of the term piedmont? (p. 541)

____

5. How is the Piedmont distinguished from the Coastal Plain? (p. 541)

____

____

6. In what way are the Tetons geologically distinctive? (p. 543)

____

7. Describe the Columbia Plateau. (p. 543)

____

____

8. What structural process determined the features of the Basin-and-Range Province? (p. 544)

____

9. Distinguish between the Cascades and the Sierra Nevada Mountains. (p. 545)

____

____

10. What is the composition of the Sierra Nevada Range? (p. 545)

____

____

(F) MATCHING

In which physiographic ***province*** is each of the following located?

_______	1. Fresno, California	a. Appalachian Plateau
_______	2. eastern Ohio	b. Basin and Range
_______	3. Las Vegas	c. Central Valley
_______	4. eastern Oregon	d. Ozark Plateau
_______	5. central Missouri	e. Columbia Plateau

(G) ESSAY QUESTIONS

1. Describe the differences between the Columbia Plateau and the Colorado Plateau.

2. Describe the changing levels of the Great Salt Lake.

(H) GRAPHICACY

1. Compare the unit opening photo on page 538 of the text with Figure 53-1.

   a. Identify and distinguish the physiographic regions on the photo by using the map as a guide.

   b. What extra regions would you put into the photo? Explain.

   c. What is the value of using both photo and sketch map for analysis?

(I) HYPOTHESIS CONSTRUCTION

1. Human use of the environment varies greatly over the surface of the earth, depending on the physical characteristics and on how we view the environment.

   a. Discuss the role of human *perception* of the environment, using examples from this and other units.

   b. Discuss the role of the physical environment as presenting us with limitations and possibilities.

   c. In a one-sentence statement summarize the arguments you have made in (a) and (b) above.

Answer key

Answers for Unit 1

Multiple Choice: 1)b; 2)c; 3)c; 4)b; 5)b
True/False: 1)F; 2)T; 3)T; 4)F; 5)T
Matching: 1)c; 2)e; 3)f; 4)g; 5)a; 6)b; 7)d

Answers for Unit 2

Multiple Choice: 1)c; 2)b; 3)a; 4)c; 5)c
True/False: 1)T; 2)T; 3)T; 4)F; 5)F
Matching: 1)c; 2)d; 3)a; 4)f; 5)g; 6)b; 7)e

Answers for Unit 3

Multiple Choice: 1)c; 2)b; 3)c; 4)b; 5)b
True/False: 1)F; 2)F; 3)T; 4)F; 5)T
Matching: 1)f; 2)i; 3) g; 4)a; 5)h; 6)b; 7)c; 8)e; 9)d

Answers for Unit 4

Multiple Choice: 1)b; 2)c; 3)a; 4)c; 5)c
True/False : 1)T; 2)F; 3)T; 4)T; 5)T
Matching: 1)e; 2)f; 3)a; 4)b; 5)c; 6)d

Answers for Unit 5

Multiple Choice: 1)c; 2)b; 3)b; 4)a; 5)a
True/False: 1)F; 2)T; 3)T; 4)F; 5)T
Matching: 1)e; 2)d; 3)g; 4)f; 5)b; 6)a; 7)c

Answers for Unit 6

Multiple Choice: 1)b; 2)a; 3)b; 4)c; 5)a
True/False: 1)T; 2)F; 3)T; 4)F; 5)T
Matching: 1)c; 2)e; 3)f; 4)a; 5)g; 6)b; 7)d

Answers for Unit 7

Multiple Choice: 1)a; 2)c; 3)b; 4)b; 5)c
True/False: 1)T; 2)T; 3)F; 4)T; 5)T
Matching: 1)b; 2)d; 3)e; 4)f; 5)a; 6)g; 7)c

Answers for Unit 8

Multiple Choice: 1)a; 2)c; 3)b; 4)b; 5)b
True/False : 1)F; 2)T; 3)T; 4)F; 5)F
Matching: 1)e; 2)d; 3)g; 4)b; 5)c; 6)a; 7)f

Answers for Unit 9

Multiple Choice: 1(b); 2 (c); 3 (c); 4 (b); 5 (a)
True/False: 1) F; 2 T; 3 F; 4 T; 5 F
Matching : 1)d ; 2)e; 3)f; 4)g; 5)c; 6)a; 7)b

Answers for Unit 10

Multiple Choice: 1)a; 2)c; 3)c; 4)b; 5)c
True/False: 1)F; 2)T; 3)F; 4)T; 5)T
Matching: 1)d; 2)e; 3)f; 4)a; 5)g; 6)c; 7)b

Answers for Unit 11

Multiple Choice: 1(c); 2(b); 3(c); 4(b); 5(b)
True/False: 1) F; 2) F; 3) T; 4) T; 5) F
Matching: 1)c; 2)e; 3)f; 4)g; 5)b; 6)a; 7)d

Answers for Unit 12

Multiple Choice: 1)b; 2)b; 3)c; 4)b; 5)a
True/False: 1)T; 2)F; 3)T; 4)F; 5)T
Matching: 1)d; 2)e; 3)f; 4)b; 5)g; 6)a; 7)c

Answers for Unit 13

Multiple Choice: 1)c; 2)a; 3)c; 4)c; 5)b
True/False: 1)T; 2)F; 3)F; 4)F; 5)T
Matching: 1)f; 2)g; 3)e; 4)b; 5)a; 6)d; 7)c

Answers for Unit 14

Multiple Choice: 1)b; 2)a; 3)c; 4)b; 5)c
True/False: 1)F; 2)T; 3)F; 4)T; 5)T
Matching: 1)d; 2)f; 3)g; 4)b; 5)a; 6)c; 7)e

Answers for Unit 15

Multiple Choice: 1)c; 2)b; 3)c; 4)b; 5)a
True/False: 1)T; 2)T; 3)F; 4)T; 5)F
Matching: 1)b; 2)c; 3)a

Answers for Unit 16

Multiple Choice: 1)b; 2)c; 3)a; 4)b; 5)a
True/False: 1)T; 2)F; 3)F; 4)T; 5)T
Matching: 1)c; 2)f; 3)e; 4)g; 5)a; 6)b; 7)d

Answers for Unit 17

Multiple Choice: 1)c; 2)a; 3)c; 4)b; 5)b
True/False: 1)T; 2)T; 3)F; 4)T; 5)T
Matching: 1)c; 2)d; 3)f; 4)a; 5)g; 6)b; 7)e

Answers for Unit 18

Multiple Choice: 1)c; 2)c; 3)b; 4)c; 5)a
True/False: 1)T; 2)F; 3)T; 4)T; 5)F
Matching: 1)d; 2)a; 3)e; 4)b; 5)c

Answers for Unit 19

Multiple Choice: 1)c; 2)b; 3)b; 4)c; 5)b
True/False: 1)T; 2)F; 3)F; 4)T; 5)T
Matching: 1)e; 2)g; 3)d; 4)f; 5)c; 6)a; 7)b

Answers for Unit 20

Multiple Choice: 1)c; 2)b; 3)c; 4)b; 5)a
True/False: 1)F; 2)T; 3)F; 4)T; 5)T
Matching: 1)e; 2)d; 3)g; 4)f; 5)c; 6)a; 7)b

Answers for Unit 21

Multiple Choice: 1)a; 2)c; 3)a; 4)b; 5)c
True/False: 1)T; 2)F; 3)T; 4)F; 5)T
Matching: 1)d; 2)e; 3)a; 4)b; 5)c

Answers for Unit 22

Multiple Choice: 1)b; 2)c; 3)a; 4)c; 5)b
True/False: 1)T; 2)F; 3)T; 4)F; 5)T
Matching: 1)e; 2)d; 3)f; 4)a; 5)b; 6)c

Answers for Unit 23

Multiple Choice: 1)c; 2)b; 3)a; 4)c; 5)b
True/False: 1)T; 2)F; 3)T; 4)F; 5)T
Matching: 1)c; 2)d; 3)f; 4)g; 5)a; 6)b; 7)e

Answers for Unit 24

Multiple Choice: 1)b; 2)c; 3)a; 4)a; 5)b
True/False: 1)F; 2)F; 3)T; 4)F; 5)F
Matching: 1)d; 2)b; 3)e; 4)f; 5)g; 6)c; 7)a

Answers for Unit 25

Multiple Choice: 1)c; 2)b; 3)c; 4)a; 5)a
True/False: 1)T; 2)F; 3)T; 4)F; 5)T
Matching: 1)f; 2)d; 3)g; 4)b; 5)a; 6)c; 7)e

Answers for Unit 26

Multiple Choice: 1)b; 2)b; 3)a; 4)c; 5)a
True/False: 1)F; 2)T; 3)F; 4)T; 5)F
Matching: 1)e; 2)a; 3)f; 4)c; 5)g; 6)d; 7)b

Answers for Unit 27

Multiple Choice: 1)c; 2)b; 3)a; 4)c; 5)b
True/False: 1)T; 2)T; 3)F; 4)T; 5)F
Matching: 1)c; 2)e; 3)g; 4)f; 5)b; 6)a; 7)d

Answers for Unit 28

Multiple Choice: 1)b; 2)c; 3)c; 4)b; 5)c
True/False: 1)F; 2)T; 3)F; 4)T; 5)F
Matching: 1)b; 2)d; 3)e; 4)a; 5)f; 6)c

Answers for Unit 29

Multiple Choice: 1)c; 2)b; 3)c; 4)a; 5)c
True/False: 1)T; 2)F; 3)T; 4)F; 5)T
Matching: 1)g; 2)f; 3)e; 4); 5)c; 6)f; 7)b

Answers for Unit 30

Multiple Choice: 1)c; 2)a; 3)b; 4)a; 5)c
True/False: 1)T; 2)F; 3)T; 4)T; 5)F
Matching: 1)f; 2)d; 3)g; 4)a; 5)b; 6)c; 7)e

Answers for Unit 31

Multiple Choice: 1)c; 2)a; 3)c; 4)a; 5)b
True/False: 1)T; 2)T; 3)F; 4)T; 5)F
Matching: 1)g; 2)d; 3)f; 4)e; 5)b; 6)c; 7) a

Answers for Unit 32

Multiple Choice: 1)b; 2)a; 3)b; 4)a; 5)c
True/False: 1)T; 2)F; 3)F; 4)T; 5)F
Matching: 1)d; 2)a; 3)f; 4)g; 5)c; 6)b; 7) e

Answers for Unit 33

Multiple Choice: 1)c; 2)c; 3)b; 4)b; 5)a
True/False: 1)F; 2)T; 3)F; 4)T; 5)T
Matching: 1)d; 2)c; 3)a; 4)b

Answers for Unit 34

Multiple Choice: 1)c; 2)b; 3)c; 4)b; 5)a
True/False: 1)T; 2)F; 3)T; 4)F; 5)F
Matching: 1)e; 2)g; 3)d; 4)f; 5)a; 6)b; 7)c

Answers for Unit 35

Multiple Choice: 1)c; 2)a; 3)b; 4)a; 5)b
True/False: 1)T; 2)F; 3)F; 4)T; 5)F
Matching: 1)f; 2)g; 3)e; 4)c; 5)a; 6)d; 7)b

Answers for Unit 36

Multiple Choice: 1)b; 2)c; 3)b; 4)a; 5)c
True/False: 1)T; 2)F; 3)T; 4)F; 5)T
Matching: 1)e; 2)g; 3)f; 4)a; 5)b; 6)d; 7)c

Answers for Unit 37

Multiple Choice: 1)c; 2)b; 3)a; 4)c; 5)b
True/False: 1)F; 2)T; 3)T; 4)F; 5)T
Matching: 1)e; 2)d; 3)b; 4)a; 5)f; 6)g; 7)c

Answers for Unit 38

Multiple Choice: 1)b; 2)c; 3)a; 4)a; 5)b
True/False: 1)T; 2)F; 3)F; 4)T; 5)T
Matching: 1)c; 2)d; 3)a; 4)b

Answers for Unit 39

Multiple Choice: 1)a; 2)c; 3)b; 4)a; 5)c
True/False: 1)T; 2)T; 3)F; 4)F; 5)F
Matching: 1)d; 2) c; 3)a; 4) b

Answers for Unit 40

Multiple Choice: 1)c; 2)b; 3)a; 4)b; 5)c
True/False: 1)T; 2)F; 3)F; 4)T; 5)F
Matching: 1)e; 2)f; 3)g; 4)a; 5)c; 6)b; 7)d

Answers for Unit 41

Multiple Choice: 1)c; 2)a; 3)c; 4)b; 5)c
True/False: 1)F; 2)F; 3)T; 4)T; 5)T
Matching: 1)c; 2)g; 3)f; 4)a; 5)b; 6)e; 7)d

Answers for Unit 42

Multiple Choice: 1)a; 2)c; 3)c; 4)b; 5)a
True/False: 1)T; 2)T; 3)F; 4)T; 5)T
Matching: 1)c; 2)f; 3)e; 4)g; 5)b; 6)d; 7)a

Answers for Unit 43

Multiple Choice: 1)a; 2)b; 3)c; 4)b; 5)a
True/False: 1)T; 2)T; 3)T; 4)F; 5)F
Matching: 1)c; 2)f; 3)g; 4)b; 5)a; 6)d; 7)e

Answers for Unit 44

Multiple Choice: 1)c; 2)b; 3)c; 4)b; 5)a
True/False: 1)F; 2)T; 3)T; 4)T; 5)F
Matching: 1)b; 2)g; 3)f; 4)e; 5)c; 6)a; 7)d

Answers for Unit 45

Multiple Choice: 1)c; 2)a; 3)a; 4)b; 5)c
True/False: 1)T; 2)T; 3)T; 4)F; 5)T
Matching: 1)f; 2)g; 3)d; 4)a; 5)b; 6)c; 7)e

Answers for Unit 46

Multiple Choice: 1)c; 2)b; 3)c; 4)a; 5)b
True/False: 1)F; 2)T; 3)F; 4)T; 5)F
Matching: 1)d; 2)f; 3)g; 4)c; 5)a; 6)h; 7)e; 8) b

Answers for Unit 47

Multiple Choice: 1)b; 2)b; 3)c; 4)b; 5)a
True/False: 1)F; 2)T; 3)F; 4)F; 5)T
Matching: 1)g; 2)e; 3)a; 4)f; 5)b; 6)d; 7)c

Answers for Unit 48

Multiple Choice: 1)c; 2)b; 3)b; 4)a; 5)b
True/False: 1)T; 2)F; 3)F; 4)T; 5)T
Matching: 1)e; 2)d; 3)a; 4)b; 5)c

Answers for Unit 49

Multiple Choice: 1)b; 2)a; 3)c; 4)c; 5)c
True/False: 1)F; 2)T; 3)T; 4)F; 5)T
Matching: 1)c; 2)f; 3)a; 4)g; 5)b; 6)d; 7)e

Answers for Unit 50

Multiple Choice: 1)b; 2)c; 3)a; 4)a; 5)a
True/False: 1)F; 2)T; 3)F; 4)F; 5)T
Matching: 1)f; 2)e; 3)d; 4)g; 5)b; 6)a; 7)c

Answers for Unit 51

Multiple Choice: 1)c; 2)c; 3)a; 4)b; 5)b
True/False: 1)F; 2)T; 3)T; 4)F; 5)T
Matching: 1)d; 2)f; 3)e; 4)b; 5)g; 6)c; 7)a

Answers for Unit 52

Multiple Choice: 1)c; 2)a; 3)b; 4)c; 5)a
True/False: 1)T; 2)F; 3)T; 4)F; 5)T
Matching: 1)c; 2)d; 3)e; 4)b; 5)a

Answers for Unit 53

Multiple Choice: 1)b; 2)c; 3)c; 4)b; 5)a
True/False: 1)F; 2)T; 3)T; 4)F; 5)F
Matching: 1)c; 2)a; 3)b; 4)e; 5)d

Notes

Notes

Notes

Notes

Notes

Notes

Notes

Notes